SpringerBriefs in Mathematics

SpringerBriefs in Mathematics showcases expositions in all areas of mathematics and applied mathematics. Manuscripts presenting new results or a single new result in a classical field, new field, or an emerging topic, applications, or bridges between new results and already published works, are encouraged. The series is intended for mathematicians and applied mathematicians.

More information about this series at http://www.springer.com/series/10030

Vijay Gupta · Michael Th. Rassias

Moments of Linear Positive Operators and Approximation

Springer

Vijay Gupta
Department of Mathematics
Netaji Subhas University of Technology
New Delhi, India

Michael Th. Rassias
Institute of Mathematics
University of Zürich
Zürich, Switzerland

Moscow Institute of Physics
and Technology
Dolgoprudny, Russia

Institute for Advanced Study
Program in Interdisciplinary Studies
Princeton, NJ, USA

ISSN 2191-8198   ISSN 2191-8201   (electronic)
SpringerBriefs in Mathematics
ISBN 978-3-030-19454-3   ISBN 978-3-030-19455-0   (eBook)
https://doi.org/10.1007/978-3-030-19455-0

Mathematics Subject Classification (2010): 30-XX, 32-XX, 34-XX, 35-XX, 41-XX, 46-XX, 47-XX, 49-XX

This Springer imprint is published by the registered company Springer Nature Switzerland AG
The registered company address is: Gewerbestrasse 11, 6330 Cham, Switzerland

Preface

Moments of Linear Positive Operators and Approximation deals with several problems concerning positive linear operators.

In Chaps. 1 and 2, we present—without proofs—a large collection of formulas concerning the values on power functions (moments) of some of the most well-known positive linear operators studied in approximation theory. Formulas which could be very useful for any future study devoted to the properties of these linear positive operators. Subsequently, Chap. 3 is concerned with the presentation of approximation properties of certain integral type operators.

New Delhi, India
Zürich, Switzerland

Vijay Gupta
Michael Th. Rassias

Contents

Chapter 1
Some Positive Linear Operators and Moments

In approximation theory, moments play an essential role. By the well known theorem of Korovkin, one can study the convergence of operators. Recently, in [66], the moments of some discrete and Kantorovich type operators were calculated by using the concept of moment generating functions. Also, Gupta et al. in [68] estimated central moments of certain operators using this approach.

The r-th order moment of an operator $L_n(f, x)$ is given by $L_n(e_r, x)$, where $e_r(t) = t^r, r = 0, 1, 2, \ldots$ Additionally, the r-th order central moment of the operator $L_n(f, x)$ is represented as follows

$$L_n((e_1 - e_0 x)^r, x) := L_n((t - x)^r, x).$$

In the present chapter, we deal with some of the discretely defined operators, which have proven to be important, and some of them constitute generalizations of well-known operators. We provide here different techniques of obtaining the moments of some linear positive operators of discrete type.

1.1 Bernstein Polynomials

The Bernstein polynomials (see [22]) for $f \in C[0, 1]$ are defined as follows:

$$B_n(f, x) = \sum_{k=0}^{n} p_{n,k}(x) f\left(\frac{k}{n}\right), \quad x \in [0, 1] \tag{1.1.1}$$

© The Author(s), under exclusive license to Springer Nature Switzerland AG 2019
V. Gupta and M. T. Rassias, *Moments of Linear Positive
Operators and Approximation*, SpringerBriefs in Mathematics,
https://doi.org/10.1007/978-3-030-19455-0_1

where

$$p_{n,k}(x) = \binom{n}{k} x^k (1-x)^{n-k}$$

is the Bernstein basis.

For the Bernstein operators on monomials of m-th order (m-th order moment) $B_n(e_m, x)$, $e_m = t^m$, $m \in \mathbb{N} \cup \{0\}$, we use the property

$$x(1-x) p'_{n,k}(x) = (k - nx) p_{n,k}(x)$$

for Bernstein basis, and have the following recurrence relation:

$$B_n(e_{m+1}, x) = \frac{x(1-x)}{n} B'_n(e_m, x) + x B_n(e_m, x).$$

It was observed by Lorentz [98] (see also Phillips [116, Chap. 7]) that the Bernstein polynomials can be expressed in the form of forward differences as

$$B_n(f, x) = \sum_{k=0}^{n} \binom{n}{k} \Delta_h^k f(0) x^k, \qquad (1.1.2)$$

where Δ is the forward difference operator with step size $h = 1/n$. Forward differences are related to derivatives, such that

$$\frac{\Delta^m f(x_0)}{h^m} = f^{(m)}(\eta),$$

where $\eta \in (x_0, x_m)$ and $x_m = x_0 + mh$. We consider $h = 1/n$, $x_0 = 0$ and $f(x) = x^r$ when $m \geq r$. Then

$$n^m \Delta^m f(0) = 0 \ \text{ if } \ m > r \qquad (1.1.3)$$

and

$$n^r \Delta^r f(0) = f^{(r)}(\eta) = r!.$$

Thus we see from (1.1.2) with $f(x) = x^r$ and $n \geq r$ that

$$B_n(x^r, x) = a_0 x^r + a_1 x^{r-1} + \cdots + a_{r-1} x + a_r,$$

where $a_0 = 1$ for $r = 0$ and $r = 1$ and for $r \geq 2$

$$a_0 = \binom{n}{r} \frac{r!}{n^r},$$

which justifies that Bernstein polynomials do not preserve any polynomial of degree greater than one.

Obviously for the step size $1/n$, the forward difference operators satisfy:

$$\Delta f(0) = f\left(\frac{1}{n}\right) - f(0)$$

$$\Delta^2 f(0) = f\left(\frac{2}{n}\right) - 2f\left(\frac{1}{n}\right) + f(0)$$

$$\Delta^3 f(0) = f\left(\frac{3}{n}\right) - 3f\left(\frac{2}{n}\right) + 3f\left(\frac{1}{n}\right) - f(0)$$

$$\Delta^4 f(0) = f\left(\frac{4}{n}\right) - 4f\left(\frac{3}{n}\right) + 6f\left(\frac{2}{n}\right) - 4f\left(\frac{1}{n}\right) + f(0).$$

Below we present a few moments (values of the Bernstein polynomials on the power functions), using the above forward difference properties.

For $f(t) = t^2$ we have $f(0) = 0$ and $\Delta f(0) = \frac{1}{n^2}$, using (1.1.2) and (1.1.3), we obtain

$$B_n(e_2, x) = \sum_{k=0}^{2} \binom{n}{k} \Delta_h^k f(0) x^k$$

$$= \binom{n}{0} 0 + \binom{n}{1} \frac{x}{n^2} + \binom{n}{2} \frac{2x^2}{n^2}$$

$$= \frac{x^2(n-1) + x}{n}.$$

For $f(t) = t^3$ we have $f(0) = 0$ and $\Delta f(0) = \frac{1}{n^3}$, using (1.1.2) and (1.1.3), we get

$$B_n(e_3, x) = \sum_{k=0}^{3} \binom{n}{k} \Delta_h^k f(0) x^k$$

$$= \binom{n}{0} 0 + \binom{n}{1} \frac{x}{n^3} + \binom{n}{2} \frac{6x^2}{n^3} + \binom{n}{3} \frac{6x^3}{n^3}$$

$$= \frac{x}{n^2} + \frac{3(n-1)x^2}{n^2} + \frac{(n-1)(n-2)x^3}{n^2}$$

$$= \frac{x^3(n-1)(n-2) + 3x^2(n-1) + x}{n^2}.$$

For $f(t) = t^4$ we have $f(0) = 0$ and $\Delta f(0) = \frac{1}{n^4}$, using (1.1.2) and (1.1.3), we obtain

$$B_n(e_4, x) = \sum_{k=0}^{4} \binom{n}{k} \Delta_h^k f(0) x^k$$

$$= \binom{n}{0} 0 + \binom{n}{1} \frac{x}{n^4} + \binom{n}{2} \frac{14x^2}{n^4} + \binom{n}{3} \frac{36x^3}{n^4} + \binom{n}{3} \frac{24x^4}{n^4}$$

$$= \frac{x}{n^3} + \frac{7(n-1)x^2}{n^3} + \frac{6(n-1)(n-2)x^3}{n^3} + \frac{(n-1(n-2)(n-3)x^4}{n^3}$$

$$= \frac{x^4(n-1)(n-2)(n-3) + 6x^3(n-1)(n-2) + 7x^2(n-1) + x}{n^3}.$$

Pop and Farcaş in [118] used Stirling's numbers of the second kind to obtain the moments of Bernstein polynomials. Let

$$x^{[k]} = x(x-1)\cdots(x-k+1), \quad x^{[0]} = 1.$$

Then

$$x^k = \sum_{v=1}^{k} S(k, v) x^{[v]}, \quad x \in \mathbb{R}, k \in \mathbb{N}_0,$$

where $S(k, v)$ are the Stirling's numbers of the second kind. Also these numbers satisfy the relation:

$$S(a, b) = bS(a-1, b) + S(a-1, b-1),$$

with

$$S(a, 1) = S(a, a) = 1; \ S(a, b) = 0, \ \text{if } a < b.$$

Which shows that

$$S(1, 1) = 1, S(2, 1) = S(2, 2) = 1, S(3, 1) = 1, S(3, 2) = 3, S(3, 3) = 1.$$

It was proved in [118] that m-th order moments of the Bernstein polynomials (1.1.1) satisfy the following:

$$B_n(e_m, x) = \frac{1}{n^m} \sum_{k=1}^{m} n^{[k]} S(m, k) x^k.$$

Also, if we denote the m-th order central moment of the Bernstein polynomials by

$$\mu_m^{B_n}(x) = B_n((e_1 - xe_0)^m, x) = \sum_{k=0}^{n} p_{n,k}(x) \left(\frac{k}{n} - x\right)^m,$$

then, we have the following recurrence relation (see [10]):

$$n\mu_{m+1}^{B_n}(x) = x(1-x)[(\mu_m^{B_n}(x))' + m\mu_{m-1}^{B_n}(x)].$$

Consequently for each $x \in [0, 1]$, we have

$$\mu_m^{B_n}(x) = O(n^{-[(m+1)/2]}),$$

where $[\alpha]$ denotes the integral part of α.

1.2 Baskakov Operators

The Baskakov operators (see [20]), for $f \in C[0, \infty)$ are defined as

$$V_n(f, x) = \sum_{k=0}^{\infty} v_{n,k}(x) f\left(\frac{k}{n}\right), \, x \in [0, \infty) \qquad (1.2.1)$$

where

$$v_{n,k}(x) = \binom{n+k-1}{k} \frac{x^k}{(1+x)^{n+k}}$$

is the Baskakov basis function. Using the identity

$$x(1+x)v'_{n,k}(x) = (k - nx)v_{n,k}(x),$$

we derive the following recurrence relation for moments:

$$V_n(e_{m+1}, x) = \frac{x(1+x)}{n} V'_n(e_m, x) + xV_n(e_m, x).$$

Some of the moments of Baskakov operators are given as

$V_n(e_0, x) = 1$

$V_n(e_1, x) = x$

$V_n(e_2, x) = \dfrac{x^2(n+1) + x}{n}$

$V_n(e_3, x) = \dfrac{x^3(n+1)(n+2) + 3x^2(n+1) + x}{n^2}$

$V_n(e_4, x) = \dfrac{x^4(n+1)(n+2)(n+3) + 6x^3(n+1)(n+2) + 7x^2(n+1) + x}{n^3}.$

Let the m-th order central moments of the Baskakov operators be denoted by

$$\mu_m^{V_n}(x) = V_n((e_1 - xe_0)^m, x) = \sum_{k=0}^{\infty} v_{n,k}(x) \left(\frac{k}{n} - x \right)^m.$$

Then, we obtain the following recurrence relation (see [128]):

$$n\mu_{m+1}^{V_n}(x) = x(1+x)[(\mu_m^{V_n}(x))' + m\mu_{m-1}^{V_n}(x)], \quad m \geq 2.$$

Consequently for each $x \in [0, \infty)$ and $m \geq 2$, we get

$$\mu_m^{V_n}(x) = \frac{P_{1,m}(x)}{n^{[(m+1)/2]}} + \frac{P_{2,m}(x)}{n^{[(m+1)/2]+1}} + \cdots \frac{P_{[m/2],m}(x)}{n^{m-1}},$$

where $P_{i,m}, i = 1, 2, \ldots [m/2]$ are polynomials in x of degree at most m and independent of n.

1.3   Szász-Mirakyan Operators

The Szász-Mirakyan operators (see [107, 134]), for $f \in C[0, \infty)$ are defined as

$$S_n(f, x) = \sum_{k=0}^{\infty} s_{n,k}(x) f \left(\frac{k}{n} \right), x \in [0, \infty) \tag{1.3.1}$$

where

$$s_{n,k}(x) = e^{-nx} \frac{(nx)^k}{k!}$$

is the Szász basis function.
  Using the identity

$$xs'_{n,k}(x) = (k - nx)s_{n,k}(x)$$

for the Szász basis function, we have the following recurrence relation for moments

$$S_n(e_{m+1}, x) = \frac{x}{n} S'_n(e_m, x) + x S_n(e_m, x).$$

Some of the moments of Szász operators defined by (1.3.1) are given as:

$$S_n(e_0, x) = 1$$
$$S_n(e_1, x) = x$$
$$S_n(e_2, x) = x^2 + \frac{x}{n}$$
$$S_n(e_3, x) = x^3 + \frac{3x^2}{n} + \frac{x}{n^2} \tag{1.3.2}$$
$$S_n(e_4, x) = x^4 + \frac{6x^3}{n} + \frac{7x^2}{n^2} + \frac{x}{n^3}.$$

Let the m-th order central moments of the Szász-Mirakyan operators be denoted by

$$\mu_m^{S_n}(x) = S_n((e_1 - x e_0)^m, x) = \sum_{k=0}^{\infty} s_{n,k}(x) \left(\frac{k}{n} - x \right)^m.$$

Then, we have the following recurrence relation (see [74]):

$$n\mu_{m+1}^{S_n}(x) = x[(\mu_m^{S_n}(x))' + m\mu_{m-1}^{S_n}(x)], \quad m \in \mathbb{N}.$$

Consequently, we derive that

(i) $\mu_m^{S_n}(x)$ is a polynomial in x of degree $[m/2]$.
(ii) For each $x \in [0, \infty)$, $\mu_m^{S_n}(x) = O(n^{-[(m+1)/2]})$, where $[\alpha]$ denotes the integral part of α.

1.4 Stancu Operators

For $B_m^{(\alpha)} : C[0, 1] \to C[0, 1]$, with a non-negative parameter α, Stancu in [130] considered a sequence of positive linear operators, which is defined as

$$B_m^{(\alpha)}(f, x) = \sum_{k=0}^{m} f\left(\frac{k}{m} \right) b_{m,k}^{(\alpha)}(x), \tag{1.4.1}$$

where $b_{m,k}^{(\alpha)}(x)$ is the Pólya distribution with density function given by

$$b_{m,k}^{(\alpha)}(x) = \binom{m}{k} \frac{\prod_{\nu=0}^{k-1}(x + \nu\alpha) \prod_{\mu=0}^{m-k-1}(1 - x + \mu\alpha)}{\prod_{\lambda=0}^{m-1}(1 + \lambda\alpha)}.$$

Alternatively, the basis function $b_{m,k}^{(\alpha)}(x)$ can be expressed in the following form:

$$b_{m,k}^{(\alpha)}(x) = \binom{m}{k} \frac{x^{[k,-\alpha]}(1 - x)^{[m-k,-\alpha]}}{1^{[m,-\alpha]}},$$

with

$$x^{[m,h]} = x(x - h) \cdots (x - (m - 1)h), \ x^{[0,h]} = 1.$$

Using the Vandermonde formula

$$(a + b)^{[n,h]} = \sum_{k=0}^{n} \binom{n}{k} a^{[k,h]} b^{[n-k,h]}$$

and the following identity for $i, j \in \mathbb{N}$ and $h \neq 0$

$$x^{[i+j,h]} = x^{[i,h]}(x - ih)^{[j,h]}.$$

Miclăuş [109] obtained the following values of Stancu operators on monomials:

$$B_m^{(\alpha)}(e_0, x) = 1$$

$$B_m^{(\alpha)}(e_1, x) = x$$

$$B_m^{(\alpha)}(e_2, x) = \frac{1}{\alpha + 1} \left(\frac{x(1 - x)}{m} + x(x + \alpha) \right)$$

$$B_m^{(\alpha)}(e_3, x) = x^3 + \frac{(2\alpha^2 m^2 + 3\alpha m^2 + 3m - 2)x^2(1 - x)}{m^2(1 + \alpha)(1 + 2\alpha)}$$

$$+ \frac{(2\alpha^2 m^2 + 3\alpha m + 1)x(1 - x)}{m^2(1 + \alpha)(1 + 2\alpha)}$$

$$B_m^{(\alpha)}(e_4, x) = x^4 + \frac{(6\alpha^3 m^3 + 11\alpha^2 m^3 + 6\alpha m^3 + 6m^2 - 11m + 6)x^3(1 - x)}{m^3(1 + \alpha)(1 + 2\alpha)(1 + 3\alpha)}$$

$$+ \frac{(6\alpha^3 m^3 + 11\alpha^2 m^3 + 18\alpha m^2 - 12\alpha m + 7m - 6)x^2(1 - x)}{m^3(1 + \alpha)(1 + 2\alpha)(1 + 3\alpha)}$$

$$+ \frac{(6\alpha^3 m^3 + 12\alpha^2 m^2 - \alpha^2 m + 7\alpha m - \alpha + 1)x(1 - x)}{m^3(1 + \alpha)(1 + 2\alpha)(1 + 3\alpha)}.$$

Also, in another paper [110], Miclăuş obtained a relation between Stancu operators and Bernstein polynomials as:

$$B_n^{(\alpha)}(f, x) = \frac{1}{B\left(\frac{x}{\alpha}, \frac{1-x}{\alpha}\right)} \int_0^1 t^{\frac{x}{\alpha}-1}(1 - t)^{\frac{1-x}{\alpha}-1} B_n(f, t) dt,$$

where $B_n(f, t)$ are the Bernstein polynomials.

The following special cases were indicated by Stancu in [130]:

(i) For $\alpha = 0$, we get the Bernstein polynomials, defined by (1.1.1).
(ii) For $\alpha = -1/m$ the operators (1.4.1) reduce to Lagrange interpolation polynomial corresponding to f, namely

$$B_m^{(-1/m)}(f, x) := L_m\left(f; 0, \frac{1}{m}, \frac{2}{m} \cdots \frac{m}{m}; x\right)$$

$$= \sum_{k=0}^{m} f\left(\frac{k}{m}\right) b_{m,k}^{(-1/m)}(x), \qquad (1.4.2)$$

where

$$b_{m,k}^{(-1/m)}(x) = (-1)^{m-k} \frac{m^m}{k!(m-k)!} x\left(x - \frac{1}{m}\right) \cdots \left(x - \frac{k-1}{m}\right)$$
$$\left(x - \frac{k+1}{m}\right) \cdots \left(x - \frac{m}{m}\right).$$

(iii) When $\alpha = 1/m^2$ and with the change of variable $x = \frac{n}{m}y$, where n is natural number not depending on m, then for $x = k/m$ we have $y = k/n$ and one gets

$$B_m^{(1/m^2)}(f, y) = \sum_{k=0}^{m} \binom{m}{k} \frac{\prod_{v=0}^{k-1}\left(\frac{ny}{m} + \frac{v}{m^2}\right) \prod_{\mu=0}^{m-k-1}\left(1 - \frac{ny}{m} + \frac{\mu}{m^2}\right)}{\left(1 + \frac{1}{m^2}\right)\left(1 + \frac{2}{m^2}\right) \cdots \left(1 + \frac{m-1}{m^2}\right)} f\left(\frac{k}{n}\right)$$

$$= \frac{1}{k!} \prod_{s=1}^{k-1}\left(1 - \frac{s}{m}\right) \frac{\prod_{v=0}^{k-1}\left(ny + \frac{v}{m}\right) \prod_{\mu=0}^{m-k-1}\left(1 - \frac{ny}{m} + \frac{\mu}{m^2}\right)}{\left(1 + \frac{1}{m^2}\right)\left(1 + \frac{2}{m^2}\right) \cdots \left(1 + \frac{m-1}{m^2}\right)} f\left(\frac{k}{n}\right).$$

In case $m \to \infty$, we get the Szász operators defined by (1.3.1).

(iv) When $\alpha = 1/m$, Lupaş and Lupaş [101] considered the following form of Stancu operators

$$B_m^{(1/m)}(f, x) = \frac{2(m!)}{(2m)!} \sum_{k=0}^{m} \binom{m}{k} f\left(\frac{k}{m}\right) (mx)_k (m - mx)_{m-k}. \qquad (1.4.3)$$

1.5 Stancu Operators Based on Inverse Pólya-Eggenberger Distribution

In 1970, Stancu [131] introduced a class of linear positive operators for $x \in [0, \infty)$, as follows:

$$V_n^{[\alpha]}(f, x) = \sum_{k=0}^{\infty} v_{n,k}^{(\alpha)}(x) f\left(\frac{k}{n}\right), \qquad (1.5.1)$$

where

$$v_{n,k}^{(\alpha)}(x) = \binom{n+k-1}{k} \frac{1^{[n,-\alpha]} x^{[k,-\alpha]}}{(1+x)^{[n+k,-\alpha]}}$$

with

$$s^{[n,h]} = s(s - h)(s - 2h)\ldots[s - (n - 1)h], \quad s^{[0,h]} = 1.$$

This class of operators is based on the inverse Pólya–Eggenberger distribution. As a special case, if $\alpha = 0$, these operators reduce to the Baskakov operators defined by (1.2.1). These operators can be represented in terms of the usual Baskakov operators in the following manner:

$$V_n^{[\alpha]}(f, x) = \frac{1}{B\left(\frac{x}{\alpha}, \frac{1}{\alpha}\right)} \int_0^\infty \frac{t^{\frac{x}{\alpha}-1}}{(1+t)^{\frac{1+x}{\alpha}}} V_n(f, t) dt,$$

where $V_n(f, t)$ is the Baskakov operator.

Using the moments of Baskakov operators, we mention below some of the moments:

$$V_n^{[\alpha]}(e_0, x) = 1,$$

$$V_n^{[\alpha]}(e_1, x) = \frac{x}{1 - \alpha},$$

$$V_n^{[\alpha]}(e_2, x) = \frac{x(nx + x + n\alpha - \alpha + 1)}{n(1 - \alpha)(1 - 2\alpha)},$$

$$V_n^{[\alpha]}(e_3, x) = \frac{1}{n^2(1 - \alpha)(1 - 2\alpha)(1 - 3\alpha)}\Bigg[x(3x^2n + 2x^2 + x^2n^2 + 3x + 1$$

$$+ 3xn - 3\alpha x - 2\alpha + 3\alpha n + 3\alpha n^2 x + \alpha^2 + 2\alpha^2 n^2 - 3\alpha^2 n)\Bigg],$$

$$V_n^{[\alpha]}(e_4, x) = \frac{x\begin{bmatrix} n^3x^3 + 6\alpha n^3 x^2 + 11\alpha^2 n^3 x + 6\alpha^3 n^3 + 6n^2 x^3 + 11nx^3 + 6x^3 \\ +12\alpha n^2 x^2 - 11\alpha^2 nx - 6\alpha^2 n^2 x - 6\alpha nx^2 + 6n^2 x^2 - 12\alpha x^2 \\ -12\alpha^3 n^2 + 6\alpha^3 n + 6\alpha^2 x + \alpha^2 + 18nx^2 + 18\alpha n^2 x - 13\alpha^2 n \\ +12\alpha^2 n^2 + 12x^2 + 5\alpha nx - 13\alpha x + 7\alpha n - 2\alpha + 7x + 7nx + 1 \end{bmatrix}}{n^3(1 - \alpha)(1 - 2\alpha)(1 - 3\alpha)(1 - 4\alpha)}.$$

1.6   Jain Operators

In 1972, Jain [91] proposed the following generalization of the Szász-Mirakyan operators, for $0 \le \beta < 1$ as

$$S_n^\beta(f, x) := \sum_{k=0}^\infty s_{n,k}^\beta(x) f\left(\frac{k}{n}\right), \tag{1.6.1}$$

where the Jain basis function is given by

$$s_{n,k}^{\beta}(x) = \frac{nx(nx + k\beta)^{k-1}}{k!}e^{-(nx+k\beta)}.$$

In case $\beta = 0$, these operators reduce to the Szász operators defined by (1.3.1). Obviously the Jain basis function satisfy the identity

$$\sum_{k=0}^{\infty} s_{n,k}^{\beta}(x) := \sum_{k=0}^{\infty} \frac{nx(nx + k\beta)^{k-1}}{k!}e^{-(nx+k\beta)} = 1. \qquad (1.6.2)$$

Jain in [91] provided its proof, starting with Lagrange's formula

$$\phi(z) = \phi(0) + \sum_{k=1}^{\infty} \frac{z^k}{(f(z))^k \cdot k!}\left[\frac{d^{k-1}}{du^{k-1}}[(f(u))^k \phi'(u)]\right]_{u=0}$$

and subsequently setting $\phi(z) = e^{\alpha z}$ and $f(z) = e^{\beta z}$ with $|\beta z| < 1$, to obtain

$$e^{\alpha z} = \sum_{k=0}^{\infty} \alpha(\alpha + k\beta)^{k-1}\frac{z^k e^{-\beta kz}}{k!}.$$

Finally considering $\alpha = nx$ and $z = 1$, in the above inequality, the identity (1.6.2), follows immediately.

Only the first two moments have been calculated in [91]. The third and fourth moments were calculated in [36], but there were some errors. These errors have been incorporated by Gupta and Greubel in [64].

For the operators defined by (1.6.1) the moments (see [36, 91]) are given as follows:

$$S_n^{\beta}(e_0, x) = 1, \qquad S_n^{\beta}(e_1, x) = \frac{x}{1 - \beta},$$

$$S_n^{\beta}(e_2, x) = \frac{x^2}{(1 - \beta)^2} + \frac{x}{n(1 - \beta)^3},$$

$$S_n^{\beta}(e_3, x) = \frac{x^3}{(1 - \beta)^3} + \frac{3x^2}{n(1 - \beta)^4} + \frac{(1 + 2\beta)x}{n^2(1 - \beta)^5},$$

$$S_n^{\beta}(e_4, x) = \frac{x^4}{(1 - \beta)^4} + \frac{6x^3}{n(1 - \beta)^5} + \frac{(7 + 8\beta)x^2}{n^2(1 - \beta)^6} + \frac{(6\beta^2 + 8\beta + 1)x}{n^3(1 - \beta)^7}.$$

Actually the operators defined by (1.6.1) do not constitute an approximation method. For $\beta_n \in [0, 1)$ with $\lim_{n\to\infty} \beta_n = 0$ Farcaş [36] obtained the following asymptotic formula.

$$\lim_{n\to\infty} n[S_n^{\beta}(f, x) - f(x)] = \frac{x}{2}f''(x).$$

In order to preserve the linear functions, Dogru et al. [34] considered the following form of Jain operators

$$\tilde{S}_n^\beta(f, x) := \sum_{k=0}^\infty s_{n,k}^\beta(u_n(x)) f\left(\frac{k}{n}\right), \tag{1.6.3}$$

where $u_n(x) = x(1 - \beta)$. Dogru et al. [34] obtained the following result:

Suppose $C_B[0, \infty)$ denotes the space of all continuous and bounded functions on $[0, \infty)$.

Theorem 1.1 *For every $f \in C_B[0, \infty)$ one has*

$$|\tilde{S}_n^\beta(f, x) - f(x)| \leq 2K\left(f, \frac{x}{4n(1 - \beta)^2}\right), \tag{1.6.4}$$

where the K functional is given by

$$K(f, \delta) = \inf_{g \in C_B[0,\infty)} [\|f - g\| + \delta\|g''\|].$$

Theorem 1.2 [9] *For every $f \in C_B[0, \infty)$ one has*

$$|S_n^\beta(f, x) - f(x)| \leq 4K\left(f, max(x, x^2)\delta_n^2\right) + \omega\left(f, \frac{\beta x}{1 - \beta}\right), \tag{1.6.5}$$

where

$$\delta_n = \frac{1}{2}\sqrt{\frac{\beta^2}{(1 - \beta)^2} + \frac{1}{2n(1 - \beta)^3}}.$$

It was pointed out in [34], that Theorem 1.1 provides a better approximation than Theorem 1.2.

1.7   Balázs-Szabados Operators

Balázs and Szabados in [19] considered the positive linear rational operators of Bernstein-type given by

$$R_n^{[\beta]}(f, x) = \frac{1}{(1 + n^{\beta-1}x)^n} \sum_{k=0}^n \binom{n}{k}(n^{\beta-1}x)^k f\left(\frac{k}{n^\beta}\right), \tag{1.7.1}$$

with $x \geq 0$, f defined on $[0, \infty)$ and $0 < \beta < 1$.

They [19] obtained weighted estimates and investigated the uniform convergence of the operators (1.7.1). These operators constitute a special case of the more general operators introduced by Balázs in [18] as follows:

$$R_n(f, x) = \frac{1}{(1 + a_n x)^n} \sum_{k=0}^{n} \binom{n}{k} (a_n x)^k f\left(\frac{k}{b_n}\right),$$

where a_n and b_n are suitably chosen real numbers, independent of x.

In [2] Abel and Vecchia established the complete asymptotic of the operators (1.7.1). They established the following explicit expression for the moments of the operators (1.7.1):

Let $0 < \beta < 1$ and $x \geq 0$. Then, for all integers $n > x^{1/(1-\beta)}$, the moments of $R_n^{[\beta]}$ can be represented in the following way:

$$R_n^{[\beta]}(e_r, x) = \sum_{i=0}^{r} n^{-i\beta} \sum_{k=0}^{\infty} n^{-k(1-\beta)} x^{k+r-i} T(r - i, i, k), \quad r = 0, 1, 2, \ldots$$

where the numbers $T(r, i, k)$ are defined as

$$T(r, i, k) = \sum_{j=0}^{k} (-1)^{k-j} \binom{k+r-1}{k-j} S_{r+j}^r \sigma_{r+i}^{r+j}, \quad (r, i, k = 0, 1, 2, \ldots)$$

and S_j^i, σ_j^i denote the Stirling numbers of the first and second kind respectively.

1.8 Meyer-König-Zeller operators

In 1960, Meyer-König-Zeller [108] considered a sequence of positive linear operators, which they called Bernstein power series. Four years later in 1964, Cheney and Sharma [24] proposed a slight modification of the operators of [108], now usually called the Meyer-König-Zeller operators, and defined for $x \in [0, 1)$ the following:

$$M_n(f, x) = (1 - x)^{n+1} \sum_{k=0}^{\infty} \binom{n+k}{k} x^k f\left(\frac{k}{n+k}\right), \tag{1.8.1}$$

$$M_n(f, 1) = f(1).$$

Obviously the first two moments are given by

$$M_n(e_0, x) = 1, \qquad M_n(e_1, x) = x.$$

Many researchers have dealt with estimates of r-th order moments, and especially with the important case $r = 2$. In the year 1984, Alkemade [15] was the first who succeeded in establishing the expression for the second moment in terms of hyper-geometric series as follows:

$$M_n(e_2, x) = x^2 + \frac{x(1-x)^2}{n+1} {}_2F_1(1, 2; n+1; x), \quad x \in [0, 1). \tag{1.8.2}$$

In 1995, Abel [1] presented an explicit expression for the r-th, $r \in \mathbb{N}_0$ order moments of the Meyer-König-Zeller operators in terms of Laplace integral as

$$M_n(e_r, x) = 1 + (1-x) \sum_{j=0}^{r-1} \binom{r}{j+1} \frac{(-n)^{j+1}}{j!} G_j(x, n), \tag{1.8.3}$$

where

$$G_j(x, s) = \int_0^\infty F_j(x, t) e^{-st} dt, \quad s > 0, x \in [0, 1]$$

is the Laplace transform of

$$F_j(x, t) = \log^j[x + (1-x)e^t], \quad j \in \mathbb{N}_0, x \in [0, 1], t \geq 0.$$

Very recently Gavrea-Ivan in [43] established the following finite-sum representation of the MKZ second order moment in terms of elementary functions as

$$M_n(e_2, x) = n(-1)^n \left(\frac{1-x}{x} \right)^{n-1} \left(\frac{(1-x)^2 \log(1-x)}{x} \right.$$
$$\left. +1 - 2x + \sum_{k=1}^{n-1} \left(\frac{(-1)^{k+1}}{k(k+1)} \frac{x^k}{(1-x)^k} [1 + (k-1)x] \right) \right), n = 1, 2, \ldots$$

1.9   Abel-Ivan Operators

In 1977, Jain and Pethe [92] proposed a generalization of the well-known Szász-Mirakyan operators for $f \in C[0, \infty)$, namely Jain-Pathe operators as

$$P_n^\alpha(f, x) = (1 + n\alpha)^{-\frac{x}{\alpha}} \sum_{k=0}^\infty \left(\alpha + \frac{1}{n} \right)^{-k} \frac{x^{[k, -\alpha]}}{k!} f\left(\frac{k}{n} \right), \tag{1.9.1}$$

where $x^{[k, -\alpha]} = x(x + \alpha)(x + 2\alpha) \ldots [x + (k-1)\alpha]$, $x^{[0, -\alpha]} = 1$.

In 2007, Abel and Ivan in [6] considered a slightly modified form of the operators of Jain and Pethe (1.9.1) by taking $\alpha = (nc)^{-1}$ and established complete asymptotic expansion. The operators discussed in [6] are defined as

$$U_n^c(f, x) = \sum_{k=0}^{\infty} \left(\frac{c}{1+c}\right)^{ncx} \frac{(ncx)_k}{k!(1+c)^k} f\left(\frac{k}{n}\right). \tag{1.9.2}$$

As a special case, if $c \to \infty$ the operators (1.9.2) reduce to classical Szász-Mirakyan operators. Also for $c = 1$, we obtain the Lupaş operators [100], defined by

$$U_n^1(f, x) = 2^{-nx} \sum_{v=0}^{\infty} \frac{(nx)^v}{2^v \cdot v!} f\left(\frac{v}{n}\right), x \geq 0. \tag{1.9.3}$$

Now, using the binomial series

$$\sum_{k=0}^{\infty} \frac{(c)_k}{k!} z^k = (1 - z)^{-c}, |z| < 1,$$

we get

$$U_n^c(e^{\theta t}, x) = \left(\frac{c}{1+c}\right)^{ncx} \left(1 - \frac{e^{\frac{\theta}{n}}}{1+c}\right)^{-ncx}$$

$$= c^{ncx} \left(1 + c - e^{\frac{\theta}{n}}\right)^{-ncx}. \tag{1.9.4}$$

We observe that $U_n^c(e^{\theta t}, x)$ may be treated as m.g.f. of the operators U_n^c, which may be utilized to obtain the moments of (1.9.2). If we take $e_r(t) = t^r$, $r \in \mathbb{N} \cup \{0\}$, then some of the moments are given by

$$U_n^c(e_r, x) = \left[\frac{\partial^r}{\partial \theta^r} U_n^c(e^{\theta t}, x)\right]_{\theta=0}$$

$$= \left[\frac{\partial^r}{\partial \theta^r} \left\{ c^{ncx} \left(1 + c - e^{\frac{\theta}{n}}\right)^{-ncx} \right\}\right]_{\theta=0}.$$

Expanding (1.9.4) in powers of θ, we have:

$$U_n^c(e^{\theta t}, x)$$

$$= 1 + x\theta + \left(\frac{x(1+c+cnx)}{nc}\right)\frac{\theta^2}{2!} + \left(\frac{x(2+3c+c^2+3cnx+3c^2nx+c^2n^2x^2)}{c^2n^2}\right)\frac{\theta^3}{3!}$$

$$+ \frac{1}{c^3n^3}\Big[x(6+12c+7c^2+c^3+11cnx+18c^2nx+7c^3nx$$

$$+ 6c^2n^2x^2+6c^3n^2x^2+c^3n^3x^3)\Big]\frac{\theta^4}{4!} + \mathcal{O}(\theta^5).$$

The first few moments of order r may be obtained by collecting the coefficients of $\frac{\theta^r}{r!}$. Some of the moments are given below:

$$U_n^c(e_0, x) = 1$$

$$U_n^c(e_1, x) = x$$

$$U_n^c(e_2, x) = \frac{x(1 + c + cnx)}{nc}$$

$$U_n^c(e_3, x) = \frac{x(2 + 3c + c^2 + 3cnx + 3c^2nx + c^2n^2x^2)}{c^2n^2}$$

$$U_n^c(e_4, x) = \frac{x(6 + 12c + 7c^2 + c^3 + 11cnx + 18c^2nx + 7c^3nx + 6c^2n^2x^2 + 6c^3n^2x^2 + c^3n^3x^3)}{c^3n^3}.$$

1.10   Modified Baskakov Operators

For a non-negative constant a (independent of n) and $x \in [0, \infty)$, Mihesan in [111] considered the following operators:

$$V_n^a(f, x) = \sum_{k=0}^{\infty} v_{n,k}^a(x) \, f\left(\frac{k}{n}\right), \qquad (1.10.1)$$

where

$$v_{n,k}^a(x) = e^{-\frac{ax}{1+x}} \frac{\sum_{i=0}^{k} \binom{k}{i}(n)_i \, a^{k-i}}{k!} \frac{x^k}{(1+x)^{n+k}}$$

is the basis function and $(n)_i = n(n+1)\cdots(n+i-1)$, $(n)_0 = 1$ denotes the rising factorial. Obviously, if $a = 0$, then these operators (1.10.1) reduce to the classical Baskakov operators, defined by (1.2.1).

Let the m-th order moment of the Baskakov operator be denoted by

$$U_{n,m}^a(x) = V_n^a(e_m, x), \, e_m(t) = t^m, \, m \in \mathbb{N} \cup \{0\}.$$

Then using the property

$$x(1+x)^a[v_{n,k}^a(x)]' = [(k - nx)(1+x) - ax]v_{n,k}^a(x),$$

we have the following recurrence relation for moments:

$$U_{n,m+1}^a(x) = \frac{x(1+x)}{n}[U_{n,m}^a(x)]' + \left[x + \frac{ax}{n(1+x)}\right]U_{n,m}^a(x).$$

An alternate approach to find the moments of the operators (1.10.1) is to consider the moment generating function. For non-negative constant $a \geqslant 0$ and for some finite number θ, we have

$$V_n^a(e^{\theta t}, x) = e^{\frac{ax}{1+x}\left(e^{\frac{\theta}{n}}-1\right)} \left(1 + x - xe^{\frac{\theta}{n}}\right)^{-n}. \tag{1.10.2}$$

The r-th order moment can be obtained as

$$V_n^a(e_r, x) = \left[\frac{\partial^r}{\partial\theta^r} V_n^a(e^{\theta t}, x)\right]_{\theta=0}$$

$$= \left[\frac{\partial^r}{\partial\theta^r} \left\{e^{\frac{ax}{1+x}\left(e^{\frac{\theta}{n}}-1\right)} \left(1 + x - xe^{\frac{\theta}{n}}\right)^{-n}\right\}\right]_{\theta=0},$$

where $e_r(t) = t^r$, $r \in \mathbb{N} \cup \{0\}$. The first few moments of order r may be obtained by collecting the coefficients of $\frac{\theta^r}{r!}$. Some of the moments are given below:

$$V_n^a(e_0, x) = 1$$

$$V_n^a(e_1, x) = x + \frac{ax}{n(1+x)}$$

$$V_n^a(e_2, x) = x^2 + \frac{2ax^2}{n(1+x)} + \frac{ax + ax^2 + a^2x^2}{n^2(1+x)^2} + \frac{x + x^2}{n}$$

$$V_n^a(e_3, x) = \frac{n^2x^3 + 3nx^3 + 2x^3 + 3nx^2 + 3x^2 + x}{n^2}$$

$$+ \frac{a^3x^3 + 3a^2x^3 + ax^3 + 3a^2x^2 + 2ax^2 + ax}{n^3(1+x)^3}$$

$$+ \frac{3ax(nx^2 + x^2 + x)}{n^2(1+x)} + \frac{3x(a^2x^2 + ax^2 + ax)}{n^2(1+x)^2}$$

$$V_n^a(e_4, x) = \frac{n^3x^4 + 6n^2x^4 + 11nx^4 + 6x^4 + 6n^2x^3 + 18nx^3 + 12x^3 + 7nx^2 + 7x^2 + x}{n^3}$$

$$+ \frac{a^4x^4 + 6a^3x^4 + 7a^2x^4 + ax^4 + 6a^3x^3 + 14a^2x^3 + 3ax^3 + 7a^2x^2 + 3ax^2 + ax}{n^4(1+x)^4}$$

$$+ \frac{4ax(n^2x^3 + 3nx^3 + 2x^3 + 3nx^2 + 3x^2 + x)}{n^3(1+x)}$$

$$+ \frac{4x(a^3x^3 + 3a^2x^3 + ax^3 + 3a^2x^2 + 2ax^2 + ax)}{n^3(1+x)^3}$$

$$+ \frac{6(a^2x^2 + ax^2 + ax)(nx^2 + x^2 + x)}{n^3(1+x)^2}.$$

1.11   Charlier Polynomials

The operators discussed in [137] are defined as

$$L_n(f; x, a) = e^{-1} \left(1 - \frac{1}{a}\right)^{(a-1)nx} \sum_{k=0}^{\infty} \frac{C_k^{(a)}(-(a-1)nx)}{k!} f\left(\frac{k}{n}\right) \quad (1.11.1)$$

where $a > 0$, $x \in [0, \infty)$ and $C_k^{(a)}$ are the Charlier polynomials, which have the generating functions of the type

$$e^t \left(1 - \frac{t}{a}\right)^x = \sum_{k=0}^{\infty} \frac{C_k^{(a)}(x)}{k!} t^k, \quad |t| < a,$$

with explicit representation

$$C_k^{(a)}(u) = \sum_{r=0}^{k} \binom{n}{r} (-u)_r \left(\frac{1}{a}\right)_r,$$

where $(\alpha)_k$ is the Pochhammer's symbol given by

$$(\alpha)_0 = 1, \quad (\alpha)_r = \alpha(\alpha + 1) \cdots (\alpha + r - 1) \quad r = 1, 2, \ldots.$$

Note that Charlier polynomials are positive if $a > 0$, $u \leq 0$.
   Few moments of the Charlier operators L_n, (see [14]) are given by

$$L_n(e_0; x, a) = 1,$$

$$L_n(e_1; x, a) = x + \frac{1}{n}$$

$$L_n(e_2; x, a) = x^2 + \frac{x}{n}\left(3 + \frac{1}{a-1}\right) + \frac{2}{n^2}$$

$$L_n(e_3; x, a) = x^3 + \frac{x^2}{n}\left(6 + \frac{3}{a-1}\right) + \frac{x}{n^2}\left(10 + \frac{6}{a-1} + \frac{2}{(a-1)^2}\right) + \frac{5}{n^3}$$

$$L_n(e_4; x, a) = x^4 + \frac{x^3}{n}\left(10 + \frac{3}{a-1}\right) + \frac{x^2}{n^2}\left(31 + \frac{30}{a-1} + \frac{5}{(a-1)^2}\right)$$

$$+ \frac{x}{n^3}\left(37 + \frac{31}{a-1} + \frac{20}{(a-1)^2} + \frac{15}{(a-1)^3}\right) + \frac{15}{n^4}.$$

1.12 Jakimovski-Leviatan Operators

Jakimovski and Leviatan [93] gave a generalization of Szász operators by using the Appell polynomials. Let $g(z) = \sum_{k=0}^{\infty} a_k z^k$ be an analytic function in the disc $|z| < R,\ (R > 1)$ and $g(1) \neq 0$. It is well known that the Appell polynomials $p_k(x)$ are defined by the following generating functions

$$g(u)e^{ux} = \sum_{k=0}^{\infty} p_k(x)u^k. \qquad (1.12.1)$$

The Jakimovski-Leviatan [93] operators are defined as follows:

$$P_n(f, x) = \frac{e^{-nx}}{g(1)} \sum_{k=0}^{\infty} p_k(nx) f\left(\frac{k}{n}\right),$$

where $p_k(x)$ are Appell polynomials given by (1.12.1).

For the special case $g(z) = 1$ the operators P_n reduce to Szász operators. From (1.12.1), by simple computations, we obtain

$$\sum_{k=0}^{\infty} p_k(nx)k = e^{nx}\left[nxg(1) + g'(1)\right]$$

$$\sum_{k=0}^{\infty} p_k(nx)k^2 = e^{nx}\left[(n^2x^2 + nx)g(1) + (1 + 2nx)g'(1) + g''(1)\right]$$

$$\sum_{k=0}^{\infty} p_k(nx)k^3 = e^{nx}\left[(n^3x^3 + 3n^2x^2 + nx)g(1) + (3n^2x^2 + 6nx + 1)g'(1)\right.$$
$$\left. + 3(nx + 1)g''(1) + g'''(1)\right]$$

$$\sum_{k=0}^{\infty} p_k(nx)k^4 = e^{nx}\left[(n^4x^4 + 6n^3x^3 + 18n^2x^2 - 5nx)g(1)\right.$$
$$+ (4n^3x^3 + 18n^2x^2 - 4nx + 1)g'(1) + 3(6n^2x^2 + 18nx - 5)g''(1)$$
$$\left. + (4nx + 6)g'''(1) + g^{(iv)}(1)\right].$$

Using the above identities, for Jakimovski-Leviatan operators P_n, few moments are given below:

$$P_n(e_0, x) = 1, \quad P_n(e_1, x) = x + \frac{g'(1)}{ng(1)},$$

$$P_n(e_2, x) = \frac{n^2 x^2 + nx}{n^2} + \frac{(2nx + 1)}{n^2} \frac{g'(1)}{g(1)} + \frac{g''(1)}{n^2 g(1)},$$

$$P_n(e_3, x) = \left[\frac{n^3 x^3 + 3n^2 x^2 + nx}{n^3} \right] + \left[\frac{3n^2 x^2 + 6nx + 1}{n^3} \right] \frac{g'(1)}{g(1)}$$
$$+ \left[\frac{3nx + 1}{n^3} \right] \frac{g''(1)}{g(1)} + \frac{g'''(1)}{n^3 g(1)},$$

$$P_n(e_4, x) = \left[\frac{n^4 x^4 + 6n^3 x^3 + 18n^2 x^2 - 5nx}{n^4} \right] + \left[\frac{4n^3 x^3 + 18n^2 x^2 - 4nx + 1}{n^4} \right] \frac{g'(1)}{g(1)}$$
$$+ \left[\frac{3(6n^2 x^2 + 18nx - 5)}{n^4} \right] \frac{g''(1)}{g(1)} + \left[\frac{4nx + 6}{n^4} \right] \frac{g'''(1)}{n^4 g(1)} + \frac{g^{(iv)}(1)}{n^4 g(1)}.$$

1.13 Szász-Chlodowsky Type Operators

The generating functions for the Gould-Hopper polynomials are given by

$$e^{ht^{d+1}} e^{xt} = \sum_{k=0}^{\infty} g_k^{d+1}(x, h) \frac{t^k}{k!} \tag{1.13.1}$$

and with the explicit representations

$$g_k^{d+1}(x, h) = \sum_{k=0}^{[k/d+1]} \frac{k!}{s!(k - (d+1)s)!} h^s d^{k-(d+1)s},$$

where $[a]$ denotes the integer part of a.

The Szász-Chlodowsky type generalization of the Szász operators with the help of generating function (1.13.1), were defined in [21] as follows

$$G_{n,h}^{(d)}(f, x) = e^{-nx/b_n - h} \sum_{k=0}^{\infty} \frac{g_k^{d+1}(nx/b_n, h)}{k!} f\left(\frac{k}{n} b_n \right), \tag{1.13.2}$$

where $h \geq 0$ and b_n is a positive increasing sequence with the properties

$$\lim_{n \to \infty} b_n = \infty$$

and

$$\lim_{n \to \infty} \frac{b_n}{n} = 0.$$

Few moments of these operators are given below:

$$G_{n,h}^{(d)}(e_0, x) = 1, \quad G_{n,h}^{(d)}(e_1, x) = x + \frac{b_n}{n} h(d + 1),$$

$$G_{n,h}^{(d)}(e_2, x) = x^2 + \frac{b_n x}{n}(2h(d + 1) + 1) + \frac{b_n^2}{n^2} h(h + 1)(d + 1)^2,$$

$$G_{n,h}^{(d)}(e_3, x) = x^3 + \frac{3b_n x^2}{n}(h(d + 1) + 1) + \frac{3b_n^2 x}{n^2}\left(h(d + 1)(h(d + 1) + d + 2) + \frac{1}{3}\right)$$

$$+ \frac{3b_n^3}{n^3} h(d + 1)^2\left((d + 1)(h^2 + 1) + h(2d + 1)\right),$$

$$G_{n,h}^{(d)}(e_4, x) = x^4 + \frac{2b_n x^3}{n}(2h(d + 1) + 3) + \frac{6b_n^2 x^2}{n^2}\left(h^2(d + 1)^2 + h(d + 1)(d + 3) + \frac{7}{6}\right)$$

$$+ \frac{2b_n^3 x}{n^3}\left(3h^2(d + 1)^2(2d + 3) + h(d + 1)(2d^2 + 7d + 7) + 2h^3(d + 1)^3 + \frac{1}{2}\right)$$

$$+ \frac{b_n^4}{n^4}\left(h(d + 1)^4(h^3 + 6h^2 + 7h + 1)\right).$$

For Chlodowsky type q Bernstein operators, we refer the readers to [96].

1.14 Brenke Type Polynomials

Brenke type polynomials [25] have generating functions of the form

$$A(t)B(xt) = \sum_{k=0}^{\infty} p_k(x)t^k, \tag{1.14.1}$$

where A and B are analytic functions:

$$A(t) = \sum_{r=0}^{\infty} a_r t^r, a_0 \neq 0. \tag{1.14.2}$$

$$B(t) = \sum_{r=0}^{\infty} b_r t^r, b_r \neq 0 (r \geq 0) \tag{1.14.3}$$

and have the following explicit expression:

$$p_k(x) = \sum_{r=0}^{k} a_{k-r} b_r x^r, k = 0, 1, 2, \ldots$$

We shall restrict ourselves to the Brenke type polynomials satisfying:

(i) $A(1) \neq 0$, $\frac{a_{k-r}b_r}{A(1)} \geq 0$, $0 \leq r \leq k, k = 0, 1, 2, \ldots$,

(ii) $B : [0, \infty) \to (0, \infty)$,

(iii) $(1.14.1)$ and the power series $(1.14.2)$ and $(1.14.3)$ converge for $|t| < R$ $(R > 1)$.

In view of the above restrictions, Varma, Sezgin and İçöz [138] introduced the following linear positive operators including the Brenke type polynomials:

$$L_n(f; x) := \frac{1}{A(1)B(nx)} \sum_{k=0}^{\infty} p_k(nx) f\left(\frac{k}{n}\right), \qquad (1.14.4)$$

where $x \geq 0$ and $n \in \mathbb{N}$.

For all $x \in [0, \infty)$, few moments of the operators $(1.14.4)$ are given by

$$L_n(e_0; x) = 1$$

$$L_n(e_1; x) = \frac{B'(nx)}{B(nx)} x + \frac{A'(1)}{nA(1)}$$

$$L_n(e_2; x) = \frac{B''(nx)}{B(nx)} x^2 + \frac{[A(1) + 2A'(1)]B'(nx)}{nA(1)B(nx)} x + \frac{A''(1) + A'(1)}{n^2 A(1)}$$

$$L_n(e_3; x) = \frac{B'''(nx)}{B(nx)} x^3 + \frac{3[A'(1) + A(1)]B''(nx)}{nA(1)B(nx)} x^2$$

$$+ \frac{[3A''(1) + 6A'(1) + A(1)]B'(nx)}{n^2 A(1)B(nx)} x + \frac{A'''(1) + 3A''(1) + A'(1)}{n^3 A(1)}$$

$$L_n(e_4; x) = \frac{B^{iv}(nx)}{B(nx)} x^4 + \frac{[4A'(1) + 6A(1)]B'''(nx)}{nA(1)B(nx)} x^3$$

$$+ \frac{[6A''(1) + 18A'(1) + 7A(1)]B''(nx)}{n^2 A(1)B(nx)} x^2$$

$$+ \frac{[4A'''(1) + 18A''(1) + 14A'(1) + A(1)]B'(nx)}{n^3 A(1)B(nx)} x$$

$$+ \frac{A^{iv}(1) + 6A'''(1) + 7A''(1) + A'(1)}{n^4 A(1)}.$$

1.15 Dunkl Type Operators

The modified Szász-Mirakjan operators—introduced by Ispir and Atakut in [89]—are the following:

$$S_n(f, x) = \frac{1}{e^{a_n x}} \sum_{k=0}^{\infty} \frac{(a_n x)^k}{k!} f\left(\frac{k}{b_n}\right),$$

where $x \in [0, \infty)$, $n \in \mathbb{N}$, $\{a_n\}$ and $\{b_n\}$ are sequences of positive numbers satisfying

$$\lim_{n \to \infty} \frac{1}{b_n} = 0, \quad \frac{a_n}{b_n} = 1 + O\left(\frac{1}{b_n}\right).$$

Sucu [132] reconstructed the Szász operators by the use of a generalized exponential function defined for $k \in \mathbb{N}_0$, $\mu > -1/2$ by Rosenblum in [126] as

$$e_\mu(x) = \sum_{k=0}^{\infty} \frac{x^k}{\gamma_\mu(k)}$$

with

$$\gamma_\mu(2k) = \frac{2^{2k} k! \Gamma(k + \mu + 1/2)}{\Gamma(\mu + 1/2)}$$

$$\gamma_\mu(2k + 1) = \frac{2^{2k+1} k! \Gamma(k + \mu + 3/2)}{\Gamma(\mu + 1/2)}.$$

Also, γ_μ satisfy the following recurrence relation

$$\gamma_\mu(k + 1) = (k + 1 + 2\mu\theta_{k+1})\gamma_\mu(k),$$

with $\theta_k = 0$ if $k \in 2\mathbb{N}$ and $\theta_k = 1$ if $k \in 2\mathbb{N} + 1$.

For $x \geq 0$, $\mu \geq 0$ and $x \in C[0, \infty)$, Sucu [132] introduced the following operators

$$S_n^*(f, x) = \frac{1}{e_\mu(nx)} \sum_{k=0}^{\infty} \frac{(nx)^k}{\gamma_\mu(k)} f\left(\frac{k + 2\mu\theta_k}{n}\right). \tag{1.15.1}$$

Recently İlbey [87] generalized further the operators (1.15.1) and proposed the following form

$$S_n^*(f, x) = \frac{1}{e_\mu(a_n x)} \sum_{k=0}^{\infty} \frac{(a_n x)^k}{\gamma_\mu(k)} f\left(\frac{k + 2\mu\theta_k}{b_n}\right). \tag{1.15.2}$$

For the operators (1.15.2), few moments considered in [87] are given below

$$S_n^*(e_0, x) = 1$$

$$S_n^*(e_1, x) = \frac{a_n x}{b_n}$$

$$S_n^*(e_2, x) = \frac{a_n^2 x^2}{b_n^2} + \left[\frac{a_n}{b_n}\frac{1}{b_n} + 2\mu\frac{a_n}{b_n}\frac{1}{b_n}\frac{e_\mu(-a_n x)}{e_\mu(a_n x)}\right]x$$

$$S_n^*(e_3, x) = \frac{a_n^3 x^3}{b_n^3} + \left[\frac{3a_n^2}{b_n^2} - 2\mu \frac{a_n^2}{b_n^2} \frac{e_\mu(-a_n x)}{e_\mu(a_n x)} \right] \frac{x^2}{b_n}$$

$$+ \left[\frac{a_n}{b_n} + \frac{4\mu^2 a_n}{b_n} + \frac{4\mu a_n}{b_n} \frac{e_\mu(-a_n x)}{e_\mu(a_n x)} \right] \frac{x}{b_n^2}$$

$$S_n^*(e_4, x) = \frac{a_n^4 x^4}{b_n^4} + \left[\frac{6a_n^3}{b_n^3} + 4\mu \frac{a_n^3}{b_n^3} \frac{e_\mu(-a_n x)}{e_\mu(a_n x)} \right] \frac{x^3}{b_n}$$

$$+ \left[\frac{7a_n^2}{b_n^2} + \frac{4\mu^2 a_n^2}{b_n^2} - 8\mu \frac{a_n^2}{b_n^2} \frac{e_\mu(-a_n x)}{e_\mu(a_n x)} \right] \frac{x^2}{b_n^2}$$

$$+ \left[(1 + 12\mu^2) \frac{a_n}{b_n} + 2\mu (3 + 4\mu^2) \frac{e_\mu(-a_n x)}{e_\mu(a_n x)} \right] \frac{x}{b_n^3}.$$

Chapter 2
Integral Type Operators and Moments

Several well known operators of discrete type have been appropriately modified in order to discuss approximation properties of integral operators. Very recently Gupta-Rassias-Sinha in [69] provided a list of integral type operators of Durrmeyer type. In the present chapter, we discuss some of the integral type operators and present their moments, using different approaches.

2.1 Gamma Operators

Let f be a function defined on $[0, \infty)$, satisfying the following growth condition

$$|f(t)| \leq M e^{\beta t}, \ M > 0, \ \beta \geq 0, \ t \to \infty.$$

Then the Gamma operators considered by Zeng [141] are defined as:

$$\overline{G}_n(f, x) = \frac{1}{\Gamma(n) x^n} \int_0^\infty t^{n-1} e^{-t/x} f\left(\frac{t}{n}\right) dt. \qquad (2.1.1)$$

The following explicit representation for the r-th order moment of the operators (2.1.1) was given by Zeng [141]:

$$\overline{G}_n(e_r, x) = \frac{(n + r - 1)!}{(n - 1)! \cdot n^r} x^r.$$

© The Author(s), under exclusive license to Springer Nature Switzerland AG 2019
V. Gupta and M. T. Rassias, *Moments of Linear Positive Operators and Approximation*, SpringerBriefs in Mathematics,
https://doi.org/10.1007/978-3-030-19455-0_2

In 1967, Lupas and Müller [102] introduced a sequence of linear positive operators $\overline{H}_n : C(0, \infty) \to C(0, \infty)$, namely Gamma operators defined as

$$\overline{H}_n(f, x) = \int_0^\infty g_n(x, u) f\left(\frac{n}{u}\right) du, \tag{2.1.2}$$

where

$$g_n(x, u) = \frac{x^{n+1}}{n!} e^{-xu} u^n, \ x > 0.$$

Karsali in [95] proposed the following modification of the Gamma operators

$$\overline{K}_n(f, x) = \int_0^\infty g_{n+2}(x, u) du \int_0^\infty g_n(u, t) f(t) dt$$

$$= \frac{(2n+3) x^{n+3}}{n!(n+2)!} \int_0^\infty \frac{t^n}{(x+t)^{2n+4}} f(t) dt. \tag{2.1.3}$$

For the Gamma operators defined by (2.1.2), Karsali [95] obtained the following representation for moments:

$$\overline{K}_n(e_r, x) = \frac{(n+r)!(n-r+2)!}{n!(n+2)!} x^r.$$

2.2 Post-widder Type Operators

Rathore and Singh [124] (for related results cf. [80]) established an asymptotic formula, and deduced inverse and saturation theorems in simultaneous approximation. They considered a parameter p and defined the operators in the following way

$$\overline{P}_n^p(f, x) := \frac{1}{(n+p)!} \left(\frac{n}{x}\right)^{n+p+1} \int_0^\infty t^{n+p} e^{-\frac{nt}{x}} f(t) \, dt. \tag{2.2.1}$$

The special case $p = 0$ provides the operator considered in [140], and for $p = -1$ these operators reduce to the operators due to May [103], which preserve the linear functions. The r-th order moments given in [124], satisfy the representation:

$$\overline{P}_n(e_r, x) = \frac{(n+r+p+1)!}{(n+r+1)! \cdot n^r} x^r.$$

2.3 Rathore Operators

For $x \geq 0$, the Rathore operators [123] are defined as

$$R_n(f, x) = \frac{n^{nx}}{\Gamma(nx)} \int_0^\infty t^{nx-1} e^{-nt} f(t) dt. \qquad (2.3.1)$$

The m-th order, $m \in \mathbb{N} \cup \{0\}$ moments with $e_m(t) = t^m$ of the Rathore operators satisfy the relation:

$$R_n(e_m, x) = \frac{(nx)_m}{n^m},$$

where

$$(nx)_m = nx(nx + 1) \cdots (nx + m - 1) \quad \text{and} \quad (nx)_0 = 1$$

(see [106]).

Miheşan in [106] proved that the Lupaş operators U_n^1 (1.9.3) is a composition of the Rathore operators R_n (2.3.1) and the well known Szász-Mirakyan operators S_n (1.3.1) i.e.

$$U_n^1(f, x) = (R_n \circ S_n)(f, x).$$

2.4 Ismail-May Operators

The Ismail-May operators (see [90, (3.16)]) are defined as follows

$$T_n(f, x) = e^{-n\sqrt{x}} \left\{ n \int_0^\infty e^{-nt/\sqrt{x}} t^{-1/2} I_1(2n\sqrt{t}) f(t) dt + f(0) \right\}, \qquad (2.4.1)$$

where I_1 stands for the modified Bessel function of the first kind given by

$$I_n(z) = \sum_{k=0}^\infty \frac{\left(\frac{z}{2}\right)^{n+2k}}{k! \Gamma(n + k + 1)}.$$

These operators are exponential type operators and we observe that the moments satisfy the following recurrence relation:

$$T_n(e_{r+1}, x) = x T_n(e_r, x) + \frac{2x^{3/2}}{n} T_n'(e_r, x).$$

Some of the moments are given below

$$T_n(e_0, x) = 1$$
$$T_n(e_1, x) = x$$
$$T_n(e_2, x) = x^2 + \frac{2x^{3/2}}{n}$$
$$T_n(e_3, x) = x^3 + \frac{6x^{5/2}}{n} + \frac{6x^2}{n^2}$$
$$T_n(e_4, x) = x^4 + \frac{12x^{7/2}}{n} + \frac{36x^3}{n^2} + \frac{24x^{5/2}}{n^3}.$$

2.5 Stancu-Beta Operators

In 1995, Stancu [133] defined Beta operators based on the Beta function of the second kind, which for $x \in (0, \infty)$ are defined as follows:

$$\overline{A}_n(f, x) = \frac{1}{B(nx, n+1)} \int_0^\infty \frac{t^{nx-1}}{(1+t)^{nx+n+1}} f(t)dt, \qquad (2.5.1)$$

and $\overline{A}_n(f, x) = f(0)$, if $x = 0$.

In [3] Abel and Gupta obtained an estimate on the rate of convergence by means of the decomposition technique of functions of bounded variation. The r-th order moments satisfy the following representation:

$$\overline{A}_n(e_r, x) = \frac{(n+r)!}{n!} \frac{\Gamma(nx+r)}{\Gamma(nx)}.$$

Also, while studying the complex case of Stancu Beta operators, Gal and Gupta [40] proved the following:

$$\overline{A}_n(e_{r+1}, x) = \frac{nx+r}{(n-r)} \overline{A}_n(e_r, x).$$

2.6 Beta Operators of the First Kind

In order to approximate Lebesgue integrable functions f on $(0, 1)$, Khan [97] introduced the Beta operators $\overline{C}_n$ defined as follows:

$$\overline{C}_n(f, x) = \frac{1}{B(nx, n(1-x))} \int_0^1 t^{nx-1}(1-t)^{n(1-x)-1} f(t)\, dt. \qquad (2.6.1)$$

If $x = 0$, then $\overline{C}_n(f, x) = f(0)$. A slightly different form of Beta operators was considered by Lupas [99]. Abel et al. [4] estimated the rate of convergence and complete asymptotic expansion for the operators (2.6.1). The m-th order moments satisfy the following representation:

$$\overline{C}_n(e_m, x) = \frac{(nx)^{\overline{m}}}{(n)^{\overline{m}}},$$

where

$$(n)^{\overline{k}} = n(n + 1) \cdots (n + k - 1), \ (n)^{\overline{0}} = 1.$$

In [4] it was also shown that for $m = 0, 1, 2, \ldots$ the moments of the operators (2.6.1) satisfy the following relation:

$$\overline{C}_n(e_m, x)$$
$$= \sum_{k=0}^{m} \frac{1}{n^{\overline{k}}} \sum_{j=0}^{k} \begin{bmatrix} m \\ m - j \end{bmatrix} x^{m-j} \sum_{i=0}^{k-j} \left\{ \begin{matrix} m - j \\ m - j - i \end{matrix} \right\} \begin{pmatrix} m - j - i \\ k - j - i \end{pmatrix} (1 - m)^{k-j-i},$$

where

$$\begin{bmatrix} j \\ i \end{bmatrix} \quad \text{and} \quad \left\{ \begin{matrix} j \\ i \end{matrix} \right\}$$

denote the Stirling number of the first kind and Stirling number of second kind respectively, defined as

$$x^{\underline{j}} = \sum_{i=0}^{j} (-1)^{j-i} \begin{bmatrix} j \\ i \end{bmatrix} x^i, \ x^j = \sum_{i=0}^{j} \left\{ \begin{matrix} j \\ i \end{matrix} \right\} x^{\underline{i}}$$

and

$$x^{\underline{i}} = x(x - 1) \cdots (x - i + 1), \ x^{\underline{0}} = 1$$

is the falling factorial.

In 2013, Gal and Gupta [41] studied Beta operators of the first kind in strips of compact disks and proved the following recurrence relation for moments:

$$\overline{C}_n(e_{m+1}, x) = \frac{(nx + m)}{(n + m)} \overline{C}_n(e_m, x).$$

2.7 Bernstein-Durrmeyer Operators

In 1967 Durrmeyer [35] proposed the following integral modification of Bernstein polynomials:

$$\overline{B}_n(f, x) = (n+1) \sum_{k=0}^{n} p_{n,k}(x) \int_0^1 p_{n,k}(t) f(t) dt, \qquad (2.7.1)$$

where $p_{n,k}(x)$ is the Bernstein basis functions defined by (1.1.1). The r-th order $(r \in \mathbb{N})$ moment $\overline{B}_n(e_r, x)$, of Bernstein-Durrmeyer operators satisfy the following recurrence relation:

$$(n + r + 2)\overline{B}_n(e_{r+1}, x) = x(1 - x)\overline{B}_n^{(1)}(e_r, x) + (nx + r + 1)\overline{B}_n(e_r, x).$$

Next, using the identity

$$\binom{n}{k} = \frac{n!}{k!(n-k)!} = \frac{n(n-1)(n-2).....(n-k+1)}{k!} = \frac{(-1)^k(-n)_k}{k!},$$

the r-th order moment $\overline{B}_n(e_r, x)$ of the Bernstein Durrmeyer operators can also be defined for $r > -1$, as

$$\overline{B}_n(e_r, x) = \frac{\Gamma(n+2)\Gamma(r+1)}{\Gamma(n+r+2)} \, {}_2F_1(-n, -r; 1; x),$$

where the hypergeometric function is given by

$$_2F_1(a, b; c; x) = \sum_{k=0}^{\infty} \frac{(a)_k(b)_k}{(c)_k k!} x^k,$$

and $(n)_k$ is the Pochhammer symbol.

Additionally, if we denote the m-th order central moments of the Bernstein-Durrmeyer polynomials by

$$\mu_m^{\overline{B}_n}(x) = B_n((e_1 - xe_0)^m, x)$$

$$= (n + 1) \sum_{k=0}^{n} p_{n,k}(x) \int_0^1 p_{n,k}(t)(t - x)^m \, dt,$$

then Derriennic in [30], observed the following:

(i) $\mu_m^{\overline{B}_n}(x)$ is a polynomial in x of degree m.

(ii) $\mu_m^{\overline{B}_n}(x) = O(n^{-[(m+1)/2]})$, where $[\beta]$ denotes the integral part of β.

(iii) $\mu_{2m}^{\overline{B}_n}(x)$ is a polynomial in $x(1 - x)$ and is uniformly equivalent to

$$\frac{1}{n^{m+1}} \frac{(2m)!}{m!} [x(1 - x)]^m, \; n \to \infty.$$

(iv) $\mu_{2m-1}^{\overline{B}_n}(x)$ is a polynomial in $x(1-x)$ multiplied by $(1-2x)$ and is uniformly equivalent to

$$-\frac{1}{n^{m+1}}\frac{(2m)!}{2(m-1)!}(1-2x)[x(1-x)]^{m-1}, \quad n \to \infty.$$

2.8 Baskakov-Durrmeyer Operators

In 1985 Sahai and Prasad [127] proposed Baskakov-Durrmeyer operators as

$$\overline{V}_n(f,x) = (n-1)\sum_{k=0}^{\infty}v_{n,k}(x)\int_0^{\infty}v_{n,k}(t)f(t)dt, \qquad (2.8.1)$$

where $v_{n,k}(x)$ is the Baskakov basis functions defined by (1.2.1).

For $n > r+2$, the r-th order ($r \in \mathbb{N}$) moments, satisfy the following recurrence relation:

$$(n-r-2)\overline{V}_n(e_{r+1},x) = x(1+x)\overline{V}_n^{(1)}(e_r,x) + (nx+r+1)\overline{V}_n(e_r,x).$$

Alternately one can also determine the moments of Baskakov Durrmeyer operators (2.8.1) in terms of hypergeometric functions. By using the identities $k! = (1)_k$, $(r+k)! = (r+1)_k.r!$ and finally by applying Pfaff transformation one obtains

$$\overline{V}_n(e_r,x) = \frac{(n-r-2)!r!}{(n-2)!} \, {}_2F_1\left(n,-r;1;-x\right),$$

If we denote the m-th order central moments of the Baskakov-Durrmeyer operators by

$$\mu_{r,m}^{\overline{V}_n}(x) = (n-r-1)\sum_{k=0}^{\infty}v_{n+r,k}(x)\int_0^{\infty}v_{n-r,k+r}(t)(t-x)^m dt,$$

then for $n > m+r+2$, the following recurrence relation holds (see [127]):

$$(n-m-r-2)\mu_{r,m+1}^{\overline{V}_n}(x) = x(1+x)[(\mu_{r,m}^{\overline{V}_n}(x))' + 2m\mu_{r,m-1}^{\overline{V}_n}(x)]$$
$$+(m+r+1)(1+2x)\mu_{r,m}^{\overline{V}_n}(x).$$

Some other form of Baskakov-Durrmeyer type operators have been discussed in [45, 50, 84].

2.9 Szász-Durrmeyer Operators

In 1985, Mazhar and Totik [105] proposed Szász-Durrmeyer operators as

$$\overline{S}_n(f; x) = n \sum_{k=0}^{\infty} s_{n,k}(x) \int_0^{\infty} s_{n,k}(t) f(t) dt, \qquad (2.9.1)$$

where $s_{n,k}(x)$ stand for the Szász basis functions defined by (1.3.1).

For the m-th order ($m \in \mathbb{N}$) moments, we have the following recurrence relation for Szász-Durrmeyer operators:

$$n\overline{S}_n(e_{m+1}, x) = x\overline{S}_n^{(1)}(e_m, x) + (nx + m + 1)\overline{S}_n(e_m, x).$$

Alternately one can also find the moments of Szász Durrmeyer operators (2.9.1) in terms of the Kummer confluent hypergeometric function. By the use of the identities $k! = (1)_k$, $(m + k)! = (m + 1)_k m!$ and finally by applying Kummer's transformation one has

$$\overline{S}_n(e_m, x) = \frac{m!}{n^m} {}_1F_1(-m; 1; -nx),$$

where the Kummer confluent hypergeometric function is given by

$$_1F_1(a; b; x) = \sum_{k=0}^{\infty} \frac{(a)_k}{(b)_k} \frac{x^k}{k!}.$$

Below we mention an alternate approach to finding the moments of Szász Durrmeyer operators (2.9.1).

$$\overline{S}_n(f, x) = \int_0^{\infty} W_n(x, t) f(t) dt,$$

where the kernel

$$W_n(x, t) = n \sum_{k=0}^{\infty} s_{n,k}(x) s_{n,k}(t).$$

Obviously

$$e^{nx} W_n(x, t) = n \sum_{k=0}^{\infty} \frac{(nx)^k}{k!} \frac{(nt)^k}{k!} e^{-nt}.$$

Differentiating both sides with respect to x, with $D \equiv \partial/\partial x$, we get

$$e^{nx}(n + D)W_n(x, t) = n \sum_{k=0}^{\infty} k \cdot n \frac{(nx)^{k-1}}{k!} \frac{(nt)^k}{k!} e^{-nt}. \qquad (2.9.2)$$

Differentiating (2.9.2) again with respect to x on both sides, we obtain

$$e^{nx}(n + D)^2 W_n(x, t) = n \sum_{k=0}^{\infty} k(k - 1)n^2 \frac{(nx)^{k-2}}{k!} \frac{(nt)^k}{k!} e^{-nt}$$

$$= n \sum_{k=1}^{\infty} n^2 \frac{(nx)^{k-2}}{(k-1)!} \frac{(nt)^k}{(k-1)!} e^{-nt} - n \sum_{k=0}^{\infty} kn^2 \frac{(nx)^{k-2}}{k!} \frac{(nt)^k}{k!} e^{-nt}$$

$$= n \sum_{k=0}^{\infty} n^2 \frac{(nx)^{k-1}}{k!} \frac{(nt)^{k+1}}{k!} e^{-nt} - \frac{1}{x} e^{nx}(n + D)W_n(x, t)$$

$$= n \sum_{k=0}^{\infty} \frac{n^2}{x} \frac{(nx)^k}{k!} \frac{(nt)^k}{k!} t e^{-nt} - \frac{1}{x} e^{nx}(n + D)W_n(x, t)$$

implying

$$e^{nx} \left((n + D)^2 + \frac{1}{x}(n + D) \right) W_n(x, t) = n \sum_{k=0}^{\infty} \frac{n^2}{x} \frac{(nx)^k}{k!} \frac{(nt)^k}{k!}.t e^{-nt}.$$

Multiplying both sides by $e^{-nx} \dfrac{x}{n^2}$, we derive that

$$\frac{x}{n^2} \left((n + D)^2 + \frac{1}{x}(n + D) \right) W_n(x, t) = W(x, t)t. \qquad (2.9.3)$$

In general, by using (2.9.3), we get the relation

$$\overline{S}_n(e_{m+1}, x) = \frac{x}{n^2} \left((n + D)^2 + \frac{1}{x}(n + D) \right) \overline{S}_n(e_m, x).$$

Using this and the fact that $\overline{S}_n(e_0, x) = 1$, step by step iteration provides the moments of Szász Durrmeyer operators. Some of the moments are as mentioned below:

$$\overline{S}_n(e_1, x) = x + \frac{1}{n}$$

$$\overline{S}_n(e_2, x) = x^2 + \frac{4x}{n} + \frac{2}{n^2}$$

$$\overline{S}_n(e_3, x) = x^3 + \frac{9x^2}{n} + \frac{18x}{n^2} + \frac{6}{n^3}$$

$$\overline{S}_n(e_4, x) = x^4 + \frac{16x^3}{n} + \frac{72x^2}{n^2} + \frac{96x}{n^3} + \frac{24}{n^4}.$$

If we denote the m-th order central moments of the Szász-Durrmeyer operators by

$$\mu_m^{\overline{S}_n}(x) = n \sum_{k=0}^{\infty} s_{n,k}(x) \int_0^{\infty} s_{n,k}(t)(t-x)^m dt,$$

then for $m \in \mathbb{N}$, the following recurrence relation holds (see [94]):

$$n\mu_{m+1}^{\overline{S}_n}(x) = x[(\mu_m^{\overline{S}_n}(x))' + 2m\mu_{m-1}^{\overline{S}_n}(x)] + (m+1)\mu_m^{\overline{S}_n}(x).$$

A more general representation for the central moments of Szász-Durrmeyer operators was considered by Gupta in [48]. If we denote

$$\mu_{r,m}^{\overline{S}_n}(x) = n \sum_{k=0}^{\infty} s_{n,k}(x) \int_0^{\infty} s_{n,k+r}(t)(t-x)^m dt,$$

then for $m \in \mathbb{N}$, the following recurrence relation holds (see [48]):

$$n\mu_{r,m+1}^{\overline{S}_n}(x) = x[(\mu_{r,m}^{\overline{S}_n}(x))' + 2m\mu_{m-1}^{\overline{S}_n}(x)] + (r+m+1)\mu_{r,m}^{\overline{S}_n}(x).$$

Clearly

$$\mu_{r,0}^{\overline{S}_n}(x) = 1, \quad \mu_{r,1}^{\overline{S}_n}(x) = \frac{r+1}{n}, \quad \mu_{r,2}^{\overline{S}_n}(x) = \frac{2nx + (r+1)(r+2)}{n^2}.$$

2.10 Baskakov-Szász Operators

In 1993, Gupta and Srivastava [73] proposed the hybrid Durrmeyer type operators namely, Baskakov-Szász-Durrmeyer operators as

$$\overline{L}_n(f; x) = n \sum_{k=0}^{\infty} v_{n,k}(x) \int_0^{\infty} s_{n,k}(t) f(t) dt, \qquad (2.10.1)$$

where $v_{n,k}(x)$ and $s_{n,k}(t)$ are the Baskakov and Szász basis functions defined by (1.2.1) and (1.3.1) respectively.

For the m-th order ($m \in \mathbb{N}$) moments, we have the following recurrence relation:

$$n\overline{L}_n(e_{m+1}, x) = x(1+x)\overline{L}_n^{(1)}(e_m, x) + (nx + m + 1)\overline{L}_n(e_m, x).$$

Alternately one can also determine the moments of Baskakov-Szász operators (2.10.1) in terms of hypergeometric functions. By using the identities $k! = (1)_k$, $(m + k)! = (m + 1)_k.m!$ and finally by applying Pfaff transformation one has

$$\overline{L}_n(e_m, x) = \frac{m!}{n^m} {}_2F_1(n, -m; 1; -x).$$

If we denote the m-th order central moments of the Baskakov-Szász operators by

$$\mu_{r,m}^{\overline{L}_n}(x) = n \sum_{k=0}^{\infty} v_{n+r,k}(x) \int_0^{\infty} s_{n,k+r}(t)(t - x)^m dt,$$

then for $m \in \mathbb{N}$, the following recurrence relation holds (see [73]):

$$n\mu_{r,m+1}^{\overline{L}_n}(x) = x(1 + x)(\mu_{r,m}^{\overline{L}_n}(x))' + [(m + 1) + r(1 + x)]\mu_{r,m}^{\overline{L}_n}(x)$$
$$+ mx(x + 2)\mu_{r,m-1}^{\overline{L}_n}(x).$$

Obviously, $\mu_{r,0}^{\overline{L}_n}(x) = 1$ and additionally one can find from the above relation the following

$$\mu_{r,1}^{\overline{L}_n}(x) = \frac{1 + r(1 + x)}{n},$$

$$\mu_{r,2}^{\overline{L}_n}(x) = \frac{rx(1 + x) + 1 + [1 + r(1 + x)]^2 + nx(2 + x)}{n^2}.$$

Moreover, Agrawal et al. [11] proposed generalized Baskakov-Szász operators as

$$\overline{L}_{n,a}(f; x) = n \sum_{k=0}^{\infty} v_{n,k}^a(x) \int_0^{\infty} s_{n,k}(t) f(t) dt, \qquad (2.10.2)$$

where the generalized Baskakov and Szász basis functions are as defined in (1.10.1) and (1.3.1) respectively. As a special case, if $a = 0$ these operators reduce to (2.10.1). The moments of the operators (2.10.2) satisfy the following recurrence relation:

$$n(1 + x)\overline{L}_{n,a}(e_{m+1}, x)(x) = x(1 + x)^2[\overline{L}_{n,a}(e_m, x)]'$$
$$+ [(m + 1)(1 + x) + nx(1 + x) + ax]\overline{L}_{n,a}(e_{m+1}, x).$$

2.11   Szász-Baskakov Operators

Szász-Baskakov type operators were initially considered in [121] as

$$\overline{M}_n(f;x) = (n-1)\sum_{k=0}^{\infty} s_{n,k}(x) \int_0^{\infty} v_{n,k}(t)f(t)dt, \qquad (2.11.1)$$

where $s_{n,k}(x)$ and $v_{n,k}(t)$ are the Szász and Baskakov basis functions defined by (1.3.1) and (1.2.1) respectively. Prasad et al. [121] estimated some direct results, which were later improved in [46]. For the m-th order ($m \in \mathbb{N}$) moments, we have the following recurrence relation:

$$(n-m-2)\overline{M}_n(e_{m+1},x) = x\overline{M}_n^{(1)}(e_m,x) + (nx+m+1)\overline{M}_n(e_m,x).$$

Also, one can find the moments of Szász-Baskakov operators (2.11.1) in terms of the Kummer confluent hypergeometric function. By using $k! = (1)_k$, $(m+k)! = (m+1)_k \cdot m!$ and subsequently by applying Kummer's transformation one has

$$\overline{M}_n(e_m,x) = \frac{(n-m-2)!m!}{(n-2)!} \, {}_1F_1\left(-m;1;-nx\right).$$

It was also observed by Gupta and Tachev in [78] that the m-th order moments of Szász-Baskakov operators may be connected with the Laguerre polynomials $L_n^m(x)$. They used the fact

$$L_n^m(x) = \frac{(n+m)!}{n!m!} \, {}_1F_1\left(-n;m+1;x\right)$$

and showed

$$\overline{M}_n(e_m,x) = \frac{(n-m-2)!m!}{(n-2)!} L_m(-nx),$$

where $L_m(-nx) = L_m^0(-nx)$ is the simple Laguerre polynomials.

Let us denote the m-th order central moments of the Szász-Baskakov operators by

$$\mu_{r,m}^{\overline{M}_n}(x) = (n-r-1)\sum_{k=0}^{\infty} s_{n,k}(x) \int_0^{\infty} v_{n-r,k+r}(t)(t-x)^m dt,$$

then for $m \in \mathbb{N}$, the following recurrence relation holds (see [75]):

$$(n-r-m-2)\mu_{r,m+1}^{\overline{M}_n}(x) = x(\mu_{r,m}^{\overline{M}_n}(x))' + [(m+1)(1+2x) + r(1+x)]\mu_{r,m}^{\overline{M}_n}(x)$$
$$+mx(x+2)\mu_{r,m-1}^{\overline{M}_n}(x).$$

Clearly, from the above relation we have

$$\mu_{r,0}^{\overline{M}_n}(x) = 1, \quad \mu_{r,1}^{\overline{M}_n}(x) = \frac{(r+1) + x(r+2)}{(n-r-2)}.$$

Some other operators of similar type have been discussed in [38]. Moreover, in 1995, Gupta-Srivastava-Sahai [76] proposed another modification of the Szász-Mirakyan operators with weight functions of Beta basis functions, which was further generalized by Dubey-Jain in [33] and studied by Gupta-Deo in [63].

2.12 BBS Operators

In 1994, Gupta [47] proposed an integral modification of the Baskakov operators by considering the weight functions of Beta basis functions in integral form instead of usual Baskakov basis functions, and defined for $x \in [0, \infty)$, the following type of operators

$$\overline{R}_n(f, x) = \sum_{k=0}^{\infty} v_{n,k}(x) \int_0^{\infty} \hat{b}_{n,k}(t) f(t) dt, \tag{2.12.1}$$

where the Baskakov and Beta basis are respectively defined as

$$v_{n,k}(x) = \frac{(n)_k}{k!} \frac{x^k}{(1+x)^{n+k}} \quad \text{and} \quad \hat{b}_{n,k}(t) = \frac{n(n+1)_k}{k!} \frac{t^k}{(1+t)^{n+k+1}}.$$

The Pochhammer symbol $(n)_k$ is defined as

$$(n)_k = n(n+1)(n+2)(n+3) \cdots (n+k-1).$$

It was observed by Gupta in [47] that by considering the modification of the Baskakov operators in the above form (2.12.1), one may obtain a better approximation. Additionally, these operators can be expressed in alternate form as follows:

$$\overline{R}_n(f, x) = n \int_0^{\infty} f(t) \frac{1+x}{(1+x+t)^{n+1}} \, {}_2F_1\left(n+1, 1-n; 1; \frac{-xt}{1+x+t}\right) dt$$

It was shown in [85] that the moments of the operators $\overline{R}_n$, for $r > -1$ satisfy the following identity:

$$\overline{R}_n(e_r, x) = \frac{\Gamma(n-r)\Gamma(r+1)}{\Gamma(n)} (1+x)^r \, {}_2F_1\left(1-n, -r; 1; \frac{x}{1+x}\right).$$

Moreover,

$$\overline{R}_n(e_r, x) = \frac{(n+r-1)!(n-r-1)!}{((n-1)!)^2} x^r + r^2 \frac{(n+r-2)!(n-r-1)!}{((n-1)!)^2} x^{r-1} + O(n^{-2}).$$

Based on two parameters α, β satisfying the conditions $0 \leq \alpha \leq \beta$, Gupta and Yadav [85] proposed in 2012 the following Baskakov-Bera-Stancu (BBS) operators, as follows

$$\overline{R}_{n,\alpha,\beta}(f, x) = n \int_0^\infty f\left(\frac{nt+\alpha}{n+\beta}\right) \frac{1+x}{(1+x+t)^{n+1}}$$

$$_2F_1\left(n+1, 1-n; 1; \frac{-xt}{1+x+t}\right) dt. \qquad (2.12.2)$$

As a special case, if $\alpha = \beta = 0$ the operators (2.12.2) reduce to Baskakov-Beta operators (2.12.1). The moments of the BBS operators for $0 \leq \alpha \leq \beta$ as obtained in [85] satisfy the relation:

$$\overline{R}_{n,\alpha,\beta}(e_r, x) = x^r \frac{n^r}{(n+\beta)^r} \frac{(n+r-1)!(n-r-1)!}{((n-1)!)^2}$$

$$+ x^{r-1}\left\{r^2 \frac{n^r}{(n+\beta)^r} \frac{(n+r-2)!(n-r-1)!}{((n-1)!)^2}\right.$$

$$\left.+ r\alpha \frac{n^{r-1}}{(n+\beta)^r} \frac{(n+r-2)!(n-r)!}{((n-1)!)^2}\right\}$$

$$+ x^{r-2}\left\{r(r-1)^2\alpha \frac{n^{r-1}}{(n+\beta)^r} \frac{(n+r-3)!(n-r)!}{((n-1)!)^2}\right.$$

$$\left.+ \frac{r(r-1)\alpha^2}{2} \frac{n^{r-2}}{(n+\beta)^r} \frac{(n+r-3)!(n-r+1)!}{((n-1)!)^2}\right\} + O(n^{-2}).$$

2.13　Abel-Ivan-Durrmeyer Type Operators

To approximate Lebesgue integrable functions on the interval $[0, \infty)$, Gupta [52] proposed the Durrmeyer type integral modification of the operators (1.9.2) as

$$(\overline{D}_{n,c,d}f)(x) = (n-d) \sum_{v=0}^\infty p_{n,v}^{[c]}(x) \int_0^\infty b_{n,v}^{[d]}(t) f(t)dt, \quad x \geq 0 \qquad (2.13.1)$$

where

$$p_{n,v}^{[c]}(x) = \left(\frac{c}{1+c}\right)^{ncx} \frac{(ncx)_v}{v!(1+c)^v}$$

and

$$b_{n,v}^{[d]}(t) = (-1)^v \frac{t^v}{v!} \phi_{n,d}^{(v)}(t),$$

with two special cases:

1. If $\phi_{n,0}(t) = e^{-nt}$, we have $b_{n,v}^{[0]}(t) = e^{-nt} \frac{(nt)^v}{v!}$
2. If $\phi_{n,1}(t) = (1+t)^{-n}$, then we get $b_{n,v}^{[1]}(t) = \binom{n+v-1}{v} \frac{t^v}{(1+t)^{n+v}}$,

which are respectively the Szász and Baskakov basis functions. The moments of the operators $(D_{n,c,d}f)(x)$ with $e_r(x) = x^r$ are given as

$$(\overline{D}_{n,c,d}e_r)(x) = \frac{\Gamma(r+1)}{\prod_{i=1}^{r+1}(n-id)}(n-d)\left(\frac{c}{1+c}\right)^{ncx} {}_2F_1\left(ncx, r+1; 1; \frac{1}{1+c}\right).$$

Further, we have

$$(\overline{D}_{n,c,d}e_0)(x) = 1, (\overline{D}_{n,c,d}e_1)(x) = \frac{1+nx}{n-2d}$$

and

$$(\overline{D}_{n,c,d}e_2)(x) = \frac{n^2cx^2 + nx(1+4c) + 2c}{c(n-2d)(n-3d)}.$$

In [53] Gupta proposed another generalization of Abel-Ivan operators with weights of Păltănea basis functions.

2.14 Lupaş-Durrmeyer Operators

Very recently Gupta and Rassias [67] proposed the Lupaş-Durrmeyer type operators for $x \in [0, 1]$ as follows:

$$\overline{D}_n^{(1/n)}(f, x) = (n+1) \sum_{k=0}^{n} p_{n,k}^{(1/n)}(x) \int_0^1 p_{n,k}(t) f(t) dt, \qquad (2.14.1)$$

where

$$p_{n,k}^{(1/n)}(x) = \frac{2(n!)}{(2n)!}\binom{n}{k}(nx)_k(n-nx)_{n-k}$$

and

$$p_{n,k}(t) = \binom{n}{k} t^k (1-t)^{n-k}$$

are Polya and Bernstein basis functions, respectively.

Using

$$\binom{n}{k} = \frac{(-1)^k (-n)_k}{k!}, \quad (a)_{n-k} = \frac{(a)_n}{(1-a-n)_k}, \quad 0 \le k \le n,$$

$$\int_0^1 p_{n,k}(t) t^r dt = \frac{n!(k+r)!}{k!(n+r+1)!} \quad \text{and} \quad (k+r)! = (r+1)_k \cdot r!,$$

Aral and Gupta [13] obtained the following form of the moments in terms of hypergeometric functions:

$$\overline{D}_n^{(1/n)}(e_r, x) = (n+1) \sum_{k=0}^n \frac{2(n)!}{(2n)!} \frac{(-1)^k (-n)_k}{k!} (nx)_k \frac{(-1)^k (n-nx)_n}{(1-2n+nx)_k} \cdot \frac{n!(k+r)!}{k!(n+r+1)!}$$

$$= \frac{2r!(n+1)!n!(n-nx)_n}{(n+r+1)!(2n)!} \sum_{k=0}^n \frac{(-n)_k (nx)_k (r+1)_k}{(1)_k (1-2n+nx)_k} \frac{1}{k!}$$

$$= \frac{2r!(n+1)!n!(n-nx)_n}{(n+r+1)!(2n)!} \, _3F_2(-n, nx, r+1; 1, 1-2n+nx; 1)$$

$$= \frac{r!(n+1)!(n-nx)_n}{(n+r+1)!(n)_n} \, _3F_2(-n, nx, r+1; 1, 1-2n+nx; 1).$$

Some of the moments obtained in [13] are as given below:

$$\overline{D}_n^{(1/n)}(e_0, x) = 1$$

$$\overline{D}_n^{(1/n)}(e_1, x) = \frac{nx+1}{n+2}$$

$$\overline{D}_n^{(1/n)}(e_2, x) = \frac{n^3 x^2 + 5n^2 x - n^2 x^2 + 3nx + 2n + 2}{(n+1)(n+2)(n+3)}$$

$$\overline{D}_n^{(1/n)}(e_3, x) = \frac{1}{(n+2)(n+3)(n+4)} \left(n^3 x^3 + \frac{6n^4 x^2 (1-x)}{(n+1)(n+2)} \right.$$

$$\left. + \frac{6n^3 x (1-x)}{(n+1)(n+2)} + 6n^2 x^2 + \frac{12n^2 x (1-x)}{n+1} + 11nx + 6 \right)$$

$$\overline{D}_n^{(1/n)}(e_4, x) = \frac{1}{(n+2)(n+3)(n+4)(n+5)} \left(n^4 x^4 + \frac{12n^4(n^2+1)x^3(1-x)}{(n+1)(n+2)(n+3)} \right.$$

$$+ \frac{12n^4(3n-1)x^2(1-x)}{(n+1)(n+2)(n+3)} + \frac{2n^4(13n-1)x(1-x)}{n(n+1)(n+2)(n+3)}$$

$$+ 10n^3 x^3 + \frac{60n^4 x^2(1-x)}{(n+1)(n+2)} + \frac{60n^3 x(1-x)}{(n+1)(n+2)}$$

$$+ 35n^2 x^2 + \frac{70n^2 x(1-x)}{n+1} + 50nx + 24 \right).$$

Additionally, while studying the complex case of the operators (2.14.1) Gal and Gupta in [41] established the following recurrence relation for the moments:

$$\overline{D}_n^{(1/n)}(e_{m+1}, x) = \frac{[n(n-m)x + (n+m)(2m+1) + nm]}{(m+n)(m+n+2)} \overline{D}_n^{(1/n)}(e_m, x)$$

$$- \frac{m^2(m+2n-nx)}{(n+m)(n+m+1)(m+n+2)} \overline{D}_n^{(1/n)}(e_{m-1}, x), m \geq 1.$$

For integers α, β with $\alpha \leq 2$ and $\alpha - 1 \leq \beta \leq 1$, Morales-Gupta in [112] considered the general form as

$$\overline{D}_{n,\alpha,\beta}^{(1/n)}(f, x) = (n - \alpha + 1) \sum_{k=\beta^+}^{n-(\alpha-\beta)^+} p_{n,k}^{(1/n)}(x) \int_0^1 p_{n-\alpha,k-\beta}(t) f(t) dt \quad (2.14.2)$$

$$+ \beta^+ p_{n,0}^{(1/n)}(x) f(0) + (\alpha - \beta)^+ p_{n,n}^{(1/n)}(x) f(1),$$

where $a^+ = \max\{0, a\}, a \in \mathbb{R}$, and $p_{n,k}^{(1/n)}(x), p_{n,k}(t)$ are defined as in (2.14.1). Also in [112], some moments of the operators (2.14.2) have been calculated, which we present below:

$$\overline{D}_{n,\alpha,\beta}^{(1/n)}(e_0, x) = 1$$

$$\overline{D}_{n,\alpha,\beta}^{(1/n)}(e_1, x) = \frac{nx - \beta + 1}{n - \alpha + 2}$$

$$\overline{D}_{n,\alpha,\beta}^{(1/n)}(e_2, x) = \frac{n^3 x^2 + 2n^2 x - n^2 x^2 + n(n+1)x(3-2\beta) + (n+1)(1-\beta)(2-\beta)}{(n+1)(n-\alpha+2)(n-\alpha+3)}$$

$$\overline{D}_{n,\alpha,\beta}^{(1/n)}(e_3, x) = \frac{n^5 x^3 + n^4 x^2(12 - 3\beta - 3x) + n^3 x(29 - 18\beta + 3\beta^2 - 3\beta x + 2x^2)}{(n+1)(n+2)(n-\alpha+2)_3}$$

$$+ \frac{n^2(6 - 11\beta + 6\beta^2 - \beta^3 + 57x - 48\beta x + 9\beta^2 x - 12x^2 + 6\beta x^2)}{(n+1)(n+2)(n-\alpha+2)_3}$$

$$+ \frac{n(18 - 33\beta + 18\beta^2 - 3\beta^3 + 22x - 24\beta x + 6\beta^2 x) + 12 - 22\beta + 12\beta^2 - 2\beta^3}{(n+1)(n+2)(n-\alpha+2)_3}$$

and

$$\overline{D}_{n,\alpha,\beta}^{(1/n)}(e_4, x)$$

$$= \frac{n^7 x^4 + n^6 x^3 (22 - 4\beta - 6x) + n^5 x^2 (131 - 54\beta + 6\beta^2 - 36x + 11x^2)}{(n+1)(n+2)(n+3)(n-\alpha+2)_4}$$

$$+ \frac{n^4 x (206 - 154\beta + 42\beta^2 - 4\beta^3 + 222x - 168\beta x + 24\beta^2 x - 46x^2 + 28\beta x^2 - 6x^3)}{(n+1)(n+2)(n+3)(n-\alpha+2)_4}$$

$$+ \frac{n^3 (24 - 50\beta + 35\beta^2 - 10\beta^3 + \beta^4 + 828x - 792\beta x + 240\beta^2 x - 24\beta^3 x - 143x^2}{(n+1)(n+2)(n+3)(n-\alpha+2)_4}$$

$$+ \frac{n^3 (42\beta x^2 + 6\beta^2 x^2 + 60x^3 - 24\beta x^3) + n^2 (144 - 300\beta + 210\beta^2 - 60\beta^3 + 6\beta^4)}{(n+1)(n+2)(n+3)(n-\alpha+2)_4}$$

$$+ \frac{n^2 (970x - 1130\beta x + 402\beta^2 x - 44\beta^3 x - 210x^2 + 180\beta x^2 - 36\beta^2 x^2)}{(n+1)(n+2)(n+3)(n-\alpha+2)_4}$$

$$+ \frac{n(264 - 550\beta + 385\beta^2 - 110\beta^3 + 11\beta^4 + 300x - 420\beta x + 180\beta^2 x - 24\beta^3 x)}{(n+1)(n+2)(n+3)(n-\alpha+2)_4}$$

$$+ \frac{144 - 300\beta + 210\beta^2 - 60\beta^3 + 6\beta^4}{(n+1)(n+2)(n+3)(n-\alpha+2)_4}.$$

2.15 Kantorovich Operators Depending on PED and IPED

In [27], Deo et al. proposed the following general sequence of linear positive operators, based on the Pólya-Eggenberger distribution (PED) as well as on the inverse Pólya-Eggenberger distribution (IPED):

$$M_n^{(\alpha)}(f, x) = \sum_k w_{n,k}^{(\alpha)}(x) f\left(\frac{k}{n}\right), \quad x \in I, \quad n = 1, 2, ..., \tag{2.15.1}$$

where

$$w_{n,k}^{(\alpha)}(x) = \frac{n+p}{n+p+(\lambda+1)k} \binom{n+p+(\lambda+1)k}{k} \frac{\Phi_k^\alpha(x)\Phi_{n+p+\lambda k}^\alpha(1+\lambda x)}{\Phi_{n+p+(\lambda+1)k}^\alpha(1+(\lambda+1)x)}$$

and

$$\Phi_k^\alpha(x) = \prod_{i=0}^{k-1}(x + i\alpha),$$

with $0 \leqslant \alpha < 1$ (may depend only on natural number n); k, p are nonnegative integers.

Let f be a real valued continuous and bounded function on $[0, \infty)$ and $\lambda \in \{-1, 0\}$. Dhamija and Deo in [31] proposed the following Kantorovich variant of operators (2.15.1) as:

$$
V_n^{(\alpha)}(f, x) = (n + p - \lambda) \sum_k \frac{n + p}{n + p + (\lambda + 1) k} \binom{n + p + (\lambda + 1) k}{k}
$$

$$
\times \frac{\Phi_k^\alpha(x) \Phi_{n+p+\lambda k}^\alpha (1 + \lambda x)}{\Phi_{n+p+(\lambda+1)k}^\alpha (1 + (\lambda + 1) x)} \int_{\frac{k}{n+p-\lambda}}^{\frac{k+1}{n+p-\lambda}} f(t) dt
$$

$$
= (n + p - \lambda) \sum_k w_{n,k}^{(\alpha)}(x) \int_{\frac{k}{n+p-\lambda}}^{\frac{k+1}{n+p-\lambda}} f(t) dt, \quad x \in I, \qquad (2.15.2)
$$

where

$$
I = \begin{cases} [0, \infty), & \lambda = 0 \\ [0, 1], & \lambda = -1. \end{cases}
$$

For Kantorovitch operators (2.15.2), some of the moments—as calculated in [31]—are given by:

$$
V_n^{(\alpha)}(e_0, x) = 1,
$$

$$
V_n^{(\alpha)}(e_1, x) = \frac{x}{1 - (\lambda + 1) \alpha} + \frac{1 + 2\lambda x}{2(n + p - \lambda)},
$$

$$
V_n^{(\alpha)}(e_2, x) = \frac{1}{(n + p - \lambda)^2} \left[\frac{1}{3} + \frac{(n + p) x}{1 - (\lambda + 1) \alpha} \right.
$$

$$
\left. + \frac{(n + p)}{(1 - \lambda \alpha)(1 - (\lambda + 1)\alpha)} \left\{ \frac{(n + p + \lambda + 1) x (x + \alpha)}{1 - 2\alpha(\lambda + 1)} + x(1 + \lambda x) \right\} \right],
$$

$$
V_n^{(\alpha)}(e_3, x) = \frac{1}{(n + p - \lambda)^3} \left[\frac{1}{4} + \frac{2(n + p)x}{(1 - (\lambda + 1)\alpha)} \right.
$$

$$
+ \frac{3(n + p)}{2(1 - (\lambda + 1)\alpha)} \left\{ \frac{(n + p + \lambda + 1)}{(1 - \alpha\lambda)(1 - 2(\lambda + 1)\alpha)} + \frac{2(n + p + 2\lambda + 1)}{(1 - (3\lambda + 2)\alpha)} \right\} x(x + \alpha)
$$

$$
+ \frac{3(n + p)}{2(1 - \lambda\alpha)(1 - (\lambda + 1)\alpha)} x(1 + \lambda x)
$$

$$
\left. + \frac{(n + p)(n + p + 2\lambda + 1)(n + p + 2(2\lambda + 1))}{(1 - (\lambda + 1)\alpha)(1 - (3\lambda + 2)\alpha)(1 - (5\lambda + 3)\alpha)} x(x + \alpha)(x + 2\alpha) \right]
$$

and

$$
\begin{aligned}
V_n^{(\alpha)}(e_4, x) = \;& \frac{1}{(n+p-\lambda)^4}\Bigg[\frac{1}{5} + \frac{4(n+p)}{(1-(\lambda+1)\alpha)}x \\
&+ \frac{(n+p)}{(1-(\lambda+1)\alpha)}\left\{\frac{2(n+p+\lambda+1)}{(1-\lambda\alpha)(1-2(\lambda+1)\alpha)} + \frac{13(n+p+2\lambda+1)}{(1-(3\lambda+2)\alpha)}\right\}x(x+\alpha) \\
&+ \frac{2(n+p)}{(1-\lambda\alpha)(1-(\lambda+1)\alpha)}x(1+\lambda x) \\
&+ \frac{8(n+p)(n+p+2\lambda+1)(n+p+2(2\lambda+1))}{(1-(\lambda+1)\alpha)(1-(3\lambda+2)\alpha)(1-(5\lambda+3)\alpha)}x(x+\alpha)(x+2\alpha) \\
&+ \frac{(n+p)(n+p+2\lambda+1)(n+p+2(2\lambda+1))(n+p+3(2\lambda+1))}{(1-(\lambda+1)\alpha)(1-(3\lambda+2)\alpha)(1-(5\lambda+3)\alpha)(1-(7\lambda+4)\alpha)} \\
&\times x(x+\alpha)(x+2\alpha)(x+3\alpha)\Bigg].
\end{aligned}
$$

2.16   Stancu-Kantorovich Operators Based on IPED

For any bounded and integrable function f defined on $[0, +\infty)$ Deo et al. [28] introduced Stancu-Kantorovich operators based on the inverse Pólya-Eggenberger distribution, as follows

$$
K_n^{[\alpha]}(f, x) = (n-1)\sum_{k=0}^{\infty} v_{n,k}^{(\alpha)}(x) \int_{\frac{k}{n-1}}^{\frac{k+1}{n-1}} f(t)\,dt. \tag{2.16.1}
$$

where $v_{n,k}^{(\alpha)}(x)$ is the Stancu basis functions defined in (1.5.1).

It was observed in [28] that the Stancu-Kantorovich operators (2.16.1), can be represented in terms of Baskakov-Kantorovich operators in the following way

$$
K_n^{[\alpha]}(f, x) = \left(B\left(\frac{x}{\alpha}, \frac{1}{\alpha}\right)\right)^{-1} \cdot \int_0^{\infty} \frac{t^{\frac{x}{\alpha}-1}}{(1+t)^{\frac{1+x}{\alpha}}} \cdot K_n(f, t)\,dt, \tag{2.16.2}
$$

where K_n are the Baskakov-Kantorovich operators defined as

$$
K_n(f, x) = (n-1)\sum_{k=0}^{\infty}\binom{n+k-1}{k}\frac{x^k}{(1+x)^{n+k}}\int_{\frac{k}{n-1}}^{\frac{k+1}{n-1}} f(t)\,dt.
$$

The following moments were calculated in [28]:

$$
K_n^{[\alpha]}(e_0, x) = 1, \quad K_n^{[\alpha]}(e_1, x) = \frac{nx}{(n-1)(1-\alpha)} + \frac{1}{2(n-1)},
$$

$$
K_n^{[\alpha]}(e_2, x) = \frac{n^2}{(1-\alpha)(1-2\alpha)(n-1)^2}\left[x(x+\alpha) + \frac{x(x+1)}{n} + \frac{(1-3\alpha)x}{n}\right] + \frac{1}{3(n-1)^2}.
$$

2.17 Baskakov Type Pólya-Durrmeyer Operators

Inspired by the generalization of Baskakov operators by Stancu, Gupta et al. [58] proposed the Durrmeyer type modification of modified Baskakov operators in the following way:

$$\overline{V}_n^{(\alpha)}(f;x) = (n-1)\sum_{k=0}^{\infty} v_{n,k}^{(\alpha)}(x) \int_0^{\infty} v_{n,k}(t)f(t)dt, \qquad (2.17.1)$$

where $v_{n,k}^{(\alpha)}(x)$ and $v_{n,k}(t)$ are respectively the Stancu and Baskakov basis functions defined in (1.5.1) and (1.2.1) respectively. In case $\alpha = 0$ we get the Baskakov–Durrmeyer operators defined by (2.8.1). It was observed in [58] that for $\alpha > 0$ and $x \in \mathbb{R}^+$, one can write

$$\overline{V}_n^{(\alpha)}(f;x) = \frac{1}{B\left(\frac{x}{\alpha},\frac{1}{\alpha}\right)} \int_0^{\infty} \frac{t^{\frac{x}{\alpha}-1}}{(1+t)^{\frac{1+x}{\alpha}}} K_n(f;t)dt,$$

where

$$K_n(f;t) = (n-1)\sum_{k=0}^{\infty} v_{n,k}(t) \int_0^{\infty} v_{n,k}(u)f(u)du,$$

and $B(p,q)$, $p,q > 0$ is the Beta function.

Few moments as obtained in [58] are given below:

$$\overline{V}_n^{(\alpha)}(e_0;x) = 1$$

$$\overline{V}_n^{(\alpha)}(e_1;x) = \frac{nx+1-\alpha}{(1-\alpha)(n-2)}$$

$$\overline{V}_n^{(\alpha)}(e_2;x) = \frac{x(x+\alpha)n^2 + x(4+x-7\alpha)n + 2(1-2\alpha)(1-\alpha)}{(1-\alpha)(1-2\alpha)(n-2)(n-3)}$$

$$\overline{V}_n^{(\alpha)}(e_3;x) = \frac{x(\alpha+x)(2\alpha+x)n^3 + 3x(\alpha+x)(3+x-7a)n^2}{(1-\alpha)(1-2\alpha)(1-3\alpha)(n-2)(n-3)(n-4)}$$
$$+ \frac{x(-81a+9x-21xa+85a^2+18+2x^2)n + 6(1-\alpha)(1-2\alpha)(1-3\alpha)}{(1-\alpha)(1-2\alpha)(1-3\alpha)(n-2)(n-3)(n-4)}$$

$$\overline{V}_n^{(\alpha)}(e_4;x) = \frac{1}{(1-\alpha)(1-2\alpha)(1-3\alpha)(1-4\alpha)(n-2)(n-3)(n-4)(n-5)}$$
$$\cdot \left[x(x+3\alpha)(x+2\alpha)(x+\alpha)n^4 + 2x(3x+8-23\alpha)(x+2\alpha)(x+\alpha)n^3 \right.$$
$$+ x(x+\alpha)(546\alpha^2 - 137x\alpha - 408\alpha + 72 + 48x + 11x^2)n^2$$
$$+ 24(1-\alpha)(1-3\alpha)(1-2\alpha)(1-4\alpha)$$
$$- 2x(46x^2\alpha - 16x^2 + 396\alpha - 273x\alpha^2 - 3x^3 - 48$$
$$\left. - 1028\alpha^2 - 36x + 204x\alpha + 830\alpha^3)n \right].$$

Also, for $m \geq 1$, the following recurrence relation for moments was established in [58]:

$$\overline{V}_n^{(\alpha)}(e_{m+1}; x) = \frac{m^2 \left(m + 1 - \frac{x}{\alpha} - \frac{1}{\alpha} - n\right)}{\left(\frac{1}{\alpha} - m - 1\right)(n - m - 1)(n - m - 2)} \overline{V}_n^{(\alpha)}(e_{m-1}, x)$$
$$+ \frac{\left[mn + \frac{2m}{\alpha} + \frac{mx}{\alpha} + \frac{1}{\alpha} + \frac{nx}{\alpha} - 2m^2 - 3m - 1\right]}{\left(\frac{1}{\alpha} - m - 1\right)(n - m - 2)} \overline{V}_n^{(\alpha)}(e_m, x).$$

2.18   Jain-Durrmeyer Operators

The Durrmeyer type modification of the operators (1.6.1), with different weight functions, has been discussed in [61, 136]. The standard Durrmeyer type modification of the operators (1.6.1) was proposed in [64]. Actually such a modification had not been studied earlier due to its complicated form when it comes to determining moments, important to check the convergence of a linear positive operator. The operators studied in [64] are defined as follows:

$$\overline{S}_n^{\beta}(f, x) = \sum_{k=0}^{\infty} \left(\int_0^{\infty} s_{n,k}^{(\beta)}(t)\, dt\right)^{-1} s_{n,k}^{(\beta)}(x) \int_0^{\infty} s_{n,k}^{(\beta)}(t) f(t)\, dt, \quad (2.18.1)$$

where the Jain basis function $s_{n,k}^{(\beta)}$ is given in (1.6.1). As a special case, if $\beta = 0$ these operators reduce to the Szász-Mirakyan-Durrmeyer operators (2.9.1). Using Tricomi's confluent hypergeometric function and Stirling numbers of the first kind, Gupta and Greubel [64] established the moments of the operators (2.18.1) (for more details see [80]). Some of the moments are the following:

$$\overline{S}_n^{\beta}(e_0, x) = 1, \qquad \overline{S}_n^{\beta}(e_1, x) = x + \frac{1}{n(1 - \beta)},$$

$$\overline{S}_n^{\beta}(e_2, x) = x^2 + \frac{4x}{n(1 - \beta)} + \frac{2!}{n^2(1 - \beta)},$$

$$\overline{S}_n^{\beta}(e_3, x) = x^3 + \frac{9x^2}{n(1 - \beta)} + \frac{6(3 - \beta)x}{n^2(1 - \beta)^2} + \frac{3!}{n^3(1 - \beta)},$$

$$\overline{S}_n^{\beta}(e_4, x) = x^4 + \frac{16x^3}{n(1 - \beta)} + \frac{12(6 - \beta)x^2}{n^2(1 - \beta)^2} + \frac{12(3\beta^2 - 6\beta + 8)x}{n^3(1 - \beta)^3} + \frac{4!}{n^4(1 - \beta)}.$$

2.19 Generalized Bernstein-Durrmeyer Operators

For certain non-negative integers $\alpha \geq \beta \geq 0$ and certain index set $I_n \subset \{0, 1, \ldots, n\}$, the generalized Bernstein-Durrmeyer operators have the following form:

$$\overline{D}_{n,\alpha,\beta} f(x) = \sum_{k \in I_n} p_{n,k}(x) f(k/n) + (n - \alpha + 1) \sum_{k=\beta}^{n-\alpha+\beta} p_{n,k}(x) \int_0^1 p_{n-\alpha,k-\beta}(t) f(t) dt.$$

In 1987 Chen [23] as well as Goodman-Sharma [44] considered the special case $\alpha = 2$, $\beta = 1$, $I_n = \{0, n\}$, usually called the genuine Bernstein-Durrmeyer operators, which reproduce linear functions. (Here $p_{0,0}(x) = 1$ and we use the general convention $p_{n,k}(x) = 0$ if $n, k \in \mathbb{N}$ do not satisfy the condition $0 \leq k \leq n$).
In 1997 Gupta [49] studied the case $\alpha = 1$, $\beta = 0$, $I_n = \{n\}$, and later Gupta-Maheshwari [65] and Abel-Gupta-Mohapatra [5] introduced the cases $\alpha = 1$, $\beta = 1$, $I_n = \{0\}$ and $\alpha = 0$, $\beta = 1$, $I_n = \{0\}$. Also, the case $\alpha = 0$, $\beta = 0$, $I_n = \emptyset$ corresponds to the usual Bernstein-Durrmeyer operators (2.7.1).
Recently Morales-Gupta [112] considered a sequences of operators $\overline{D}_{n,\alpha,\beta}$ where $\alpha \geq \beta \geq 0$, let α and β be any elements from $\mathbb{Z}$ with $\alpha \leq 2$ and $\alpha - 1 \leq \beta \leq 1$, and

$$I_n = \{i : 0 \leq i < \beta\} \cup \{i : n - \alpha + \beta < i \leq n\}.$$

For all these values of α and β, by using the notation

$$a^+ = \max\{0, a\}, \quad a \in \mathbb{R},$$

the operators may be rewritten in the following manner:

$$\overline{D}_{n,\alpha,\beta}(f, x) = \beta^+ f(0)(1 - x)^n$$
$$+ (n - \alpha + 1) \sum_{k=\beta^+}^{n-(\alpha-\beta)^+} p_{n,k}(x) \int_0^1 p_{n-\alpha,k-\beta}(t) f(t) dt \qquad (2.19.1)$$
$$+ (\alpha - \beta)^+ f(1) x^n.$$

Notice that

$$\beta^+ = \begin{cases} 1, & \text{if } \beta = 1; \\ 0, & \text{otherwise}, \end{cases}$$

Moreover, if $\delta = \delta(t)$ denotes the Dirac delta function, one can even rewrite the operators (2.19.1) as

$$\overline{D}_{n,\alpha,\beta}(f, x) = \int_0^1 K_n^{\alpha,\beta}(x, t) f(t) dt,$$

where

$$K_n^{\alpha,\beta}(x,t) = \beta^+(1-x)^n\delta(t) + (n-\alpha+1)\sum_{k=\beta^+}^{n-(\alpha-\beta)^+} p_{n,k}(x)p_{n-\alpha,k-\beta}(t)$$
$$+ (\alpha-\beta)^+ x^n\delta(t-1).$$

Let m, n be positive integers with $n > 1$ and $m \geq 1$. Then the following form of moments of the operators (2.19.1) was proved in [112]:

$$\overline{D}_{n,\alpha,\beta}(e_m, x) = \frac{(n-\alpha+1)!}{(n-\alpha+1+m)!}\sum_{k=\beta^+}^{m}\binom{m}{k}\frac{(m-\beta)!n!}{(k-\beta)!(n-k)!}x^k.$$

From this representation, we have

$$\overline{D}_{n,\alpha,\beta}(e_1, x) = \frac{1-\beta+nx}{n-\alpha+2}$$
$$\overline{D}_{n,\alpha,\beta}(e_2, x) = \frac{(1-\beta)(2-\beta) + 2n(2-\beta)x + n(n-1)x^2}{(n+2-\alpha)(n+3-\alpha)}.$$

Consequently, for the case $\alpha = 2$, $\beta = 1$, we get genuine Bernstein polynomials, which reproduce the linear functions. Let m, n be positive integers. The following recurrence relation for central moments of $\overline{D}_{n,\alpha,\beta}$ was established in [112]:

$$\left(n+m+(2-\alpha)\right)\mu_{m+1,n}(x) = x(1-x)\left(\mu'_{m,n}(x) + 2m\mu_{m-1,n}(x)\right)$$
$$+ \left(m(1-2x) - (2-\alpha)x + (1-\beta)\right)\mu_{m,n}(x),$$

where

$$\mu_{m,n}(x) = \overline{D}_{n,\alpha,\beta}((e_1 - xe_0)^m, x).$$

2.20 Phillips Operators

In 1954, Phillips [117] proposed the following operators

$$\overline{H}_n(f, x) = \int_0^\infty W_n(x,t)f(t)dt$$
$$= n\sum_{k=1}^{\infty} s_{n,k}(x)\int_0^\infty s_{n,k-1}(t)f(t)dt + e^{-nx}f(0), \qquad (2.20.1)$$

where

$$W_n(x, t) = e^{-n(x+t)} \left(\sum_{k=1}^{\infty} \frac{(n^2 x)^k t^{k-1}}{n!(n-1)!} + \delta(t) \right)$$

with $\delta(t)$ being the Dirac delta function and $s_{n,k}(x)$ the Szász basis functions defined by (1.3.1).

These operators are called Phillips operators (see [39, 104]). We usually call these operators as genuine operators since these preserve linear functions i.e. $\overline{H}_n(at + b; x) = ax + b$.

It was observed by May in [104] that if

$$P = x \left(1 + \frac{D}{n} \right)^2 = x \left(1 + \frac{2D}{n} + \frac{D^2}{n^2} \right),$$

then

$$PW_n(x, t) = W_n(x, t)t$$

and the moments of the operators (2.20.1) can be calculated directly as follows:

$$\overline{H}_n(e_0, x) = P^0 1 = 1$$

$$\overline{H}_n(e_1, x) = P^1 1 = x \left(1 + \frac{2D}{n} + \frac{D^2}{n^2} \right).1 = x$$

$$\overline{H}_n(e_2, x) = P^2 1 = x \left(1 + \frac{2D}{n} + \frac{D^2}{n^2} \right).x = x^2 + \frac{2x}{n}$$

$$\overline{H}_n(e_3, x) = P^3 1 = x \left(1 + \frac{2D}{n} + \frac{D^2}{n^2} \right).\left(x^2 + \frac{2x}{n} \right) = x^3 + \frac{6x^2}{n} + \frac{6x}{n^2}$$

$$\overline{H}_n(e_4, x) = P^4 1 = x \left(1 + \frac{2D}{n} + \frac{D^2}{n^2} \right).\left(x^3 + \frac{6x^2}{n} + \frac{6x}{n^2} \right) = x^4 + \frac{12x^3}{n} + \frac{36x^2}{n^2} + \frac{12x}{n^3}.$$

Additionally, we may write the m-th order moments of Phillips operators in terms of a confluent hypergeometric function and connect it with the Laguerre polynomials $L_n^m(x)$ as follows

$$\overline{H}_n(e_m, x) = \frac{x\Gamma(m+1)}{n^{m-1}} {}_1F_1(1-m; 2; -nx)$$

and also, we have

$$\overline{H}_n(e_m, x) = \frac{x\Gamma(m)}{n^{m-1}} L_{n-1}^1(-nx),$$

where $L_{n-1}^1(-nx)$ stands for the generalized Laguerre polynomial.

Moreover, the moments of Phillips operators satisfy the following recurrence relation:

$$\overline{H}_n(e_{m+1}, x) = \frac{x}{n}\overline{H}_n'(e_m, x) + \left(x + \frac{m}{n} \right) \overline{H}_n(e_m, x).$$

If we denote the m-th order central moments of the Phillips operators by

$$\mu_m^{\overline{H}_n}(x) = n \sum_{k=1}^{\infty} s_{n,k}(x) \int_0^{\infty} s_{n,k-1}(t)(t-x)^m dt + e^{-nx}(-x)^m,$$

then the following recurrence relation holds (see [104]):

$$\mu_{m+1}^{\overline{H}_n}(x) = \frac{2x}{n}(\mu_m^{\overline{H}_n}(x))' + \frac{x}{n^2}(\mu_m^{\overline{H}_n}(x))''$$

$$+ \frac{2xm}{n}\mu_{m-1}^{\overline{H}_n}(x) + \frac{xm(m-1)}{n^2}\mu_{m-2}^{\overline{H}_n}(x) + \frac{2xm}{n^2}(\mu_{m-1}^{\overline{H}_n}(x))''.$$

Also, a simple proof of the recurrence relation provided in [51] is presented below:

$$n\mu_{m+1}^{\overline{H}_n}(x) = x[(\mu_m^{\overline{H}_n}(x))' + 2m\mu_{m-1}^{\overline{H}_n}(x)] + m\mu_m^{\overline{H}_n}(x).$$

In particular,

$$\mu_0^{\overline{H}_n}(x) = 1, \quad \mu_1^{\overline{H}_n}(x) = 0, \quad \mu_2^{\overline{H}_n}(x) = \frac{2x}{n}.$$

Recently, in order to generalize the Phillips operators, Păltănea in [114] proposed the following operators, which are based on certain parameter $\rho > 0$, as

$$\overline{H}_\alpha^\rho(f, x) = e^{-\alpha x} f(0) + \sum_{k=1}^{\infty} s_{\alpha,k}(x) \int_0^{\infty} \theta_{\alpha,k}^\rho(t) f(t) dt, \, x \in [0, \infty), \quad (2.20.2)$$

where the basis functions are defined as follows

$$s_{\alpha,k}(x) = e^{-\alpha x} \frac{(\alpha x)^k}{k!}, \, \theta_{\alpha,k}^\rho(t) = \frac{\alpha \rho}{\Gamma(k\rho)} e^{-\alpha \rho t} (\alpha \rho t)^{k\rho - 1}.$$

For fixed $\alpha > 0$ and $\rho > 0$, the m-th order moments of (2.20.2), for $m \in \mathbb{N} \cup \{0\}$ and $x \geq 0$, satisfy the following recurrence relation (see [114]):

$$\overline{H}_\alpha^\rho(e_m, x) = \left(x + \frac{m-1}{\alpha\rho}\right)\overline{H}_\alpha^\rho(e_{m-1}, x) + \frac{x}{\alpha}[\overline{H}_\alpha^\rho(e_{m-1}, x)]'.$$

Furthermore,

$$\overline{H}_\alpha^\rho(e_m, x) = x^m + \frac{(\rho+1)}{2\alpha\rho} \cdot m(m-1)x^{m-1}$$

$$+ \frac{(\rho+1)}{24(\alpha\rho)^2} \cdot m(m-1)(m-2)[(3m-5)\rho + 3m - 1]x^{m-2} + \cdots.$$

2.21 Genuine Baskakov-Durrmeyer Operators

The genuine Baskakov-Durrmeyer type operators discussed in [72], for $\alpha > 0$, $x \in [0, \infty)$ and $n \in \mathbb{N}$, are defined as

$$\overline{V}_{n,\alpha}(f, x) = \sum_{k=1}^{\infty} v_{n,k,\alpha}(x) \int_0^{\infty} b_{n,k,\alpha}(t) f(t) dt + (1 + \alpha x)^{-\frac{n}{\alpha}} f(0), \qquad (2.21.1)$$

where

$$v_{n,k,\alpha}(x) = \frac{\Gamma(\frac{n}{\alpha} + k)}{\Gamma(k+1)\Gamma(\frac{n}{\alpha})} \cdot \frac{(\alpha x)^k}{(1 + \alpha x)^{\frac{n}{\alpha}+k}},$$

$$b_{n,k,\alpha}(t) = \frac{\alpha \Gamma(\frac{n}{\alpha} + k + 1)}{\Gamma(k)\Gamma(\frac{n}{\alpha} + 1)} \cdot \frac{(\alpha t)^{k-1}}{(1 + \alpha t)^{\frac{n}{\alpha}+k+1}}.$$

These operators (2.21.1) reproduce constant as well as linear functions.

In the special case $\alpha = 1$, we immediately get the operators studied in [37]. An alternate form of these operators, as presented in [83], in terms of the Hypergeometric series is the following:

$$\overline{V}_{n,\alpha}(f, x) = n(n + \alpha) \int_0^{\infty} \frac{f(t)x}{(1 + \alpha t)(1 + \alpha x + \alpha t)^{\frac{n}{\alpha}+1}}$$
$$\phantom{\overline{V}_{n,\alpha}(f, x) =} {}_2F_1\left(\frac{n}{\alpha} + 1, -\frac{n}{\alpha}; 2; -\frac{\alpha^2 xt}{(1 + \alpha x + \alpha t)}\right) dt + (1 + \alpha x)^{-\frac{n}{\alpha}} f(0).$$

For $n > 0$ and $r \geq 1$, the moments of operators (2.21.1) as estimated in [83] satisfy

$$\overline{V}_{n,\alpha}(e_r, x) = \frac{x\Gamma(\frac{n}{\alpha} - r + 1)\Gamma(r + 1)}{\alpha^{r-1}\Gamma(\frac{n}{\alpha})} \, {}_2F_1\left(\frac{n}{\alpha} + 1, 1 - r; 2; -\alpha x\right).$$

For $r \geq 2$, we have

$$\overline{V}_{n,\alpha}(e_r, x) = \frac{\Gamma(\frac{n}{\alpha} + r)\Gamma(\frac{n}{\alpha} - r + 1)}{\Gamma(\frac{n}{\alpha} + 1)\Gamma(\frac{n}{\alpha})} x^r$$
$$+ \frac{x \cdot r!}{\alpha^{r-1}(\frac{n}{\alpha} - r + 1)_{r-1}} \cdot \sum_{k=0}^{r-2} \frac{(\frac{n}{\alpha} + 1)_k (1 - r)_k}{(2)_k \cdot k!} \cdot (-\alpha x)^k.$$

Additionally, if the m-th order central moments are denoted by

$$\mu_m^{\overline{V}_{n,\alpha}}(x) = \overline{V}_{n,\alpha}((t - x)^m, x),$$

then the following recurrence relation for $n > \alpha m$, estimated in [72], holds:

$$(n - \alpha m)\mu_{m+1}^{\overline{V}_{n,\alpha}}(x) = x(1 + \alpha x)[(\mu_m^{\overline{V}_{n,\alpha}}(x))' + 2m\mu_{m-1}^{\overline{V}_{n,\alpha}}(x)]$$
$$+ m(1 + 2\alpha x)\mu_m^{\overline{V}_{n,\alpha}}(x).$$

2.22 Srivastava-Gupta Operators

The Srivastava-Gupta operators (see [26, 88, 129]) are defined by

$$G_{n,c}(f, x) = n \sum_{k=1}^{\infty} p_{n,k}(x, c) \int_0^{\infty} p_{n+c,k-1}(t, c) f(t) d$$
$$+ p_{n,0}(x, c) f(0), \tag{2.22.1}$$

where

$$p_{n,k}(x, c) = \frac{(-x)^k}{k!} \phi_{n,c}^{(k)}(x)$$

with the following special cases:

- If $c = 0$ and $\phi_{n,c}(x) = e^{-nx}$ then we get

$$p_{n,k}(x, 0) = e^{-nx} \frac{(nx)^k}{k!},$$

- If $c \in \mathbb{N}$ and $\phi_{n,c}(x) = (1 + cx)^{-n/c}$, then we obtain

$$p_{n,k}(x, c) = \frac{(n/c)_k}{k!} \frac{(cx)^k}{(1 + cx)^{\frac{n}{c}+k}},$$

- If $c = -1$ and $\phi_{n,c}(x) = (1 - x)^n$ then

$$p_{n,k}(x, -1) = \binom{n}{k} x^k (1 - x)^{n-k}.$$

Here, for the last case i.e $c = -1$ we have $x \in [0, 1]$, while for $c \in \mathbb{N} \cup \{0\}$, we have $x \in [0, \infty)$. In [129] the recurrence formula has been provided to calculate the central moments. We provide below an alternate approach to calculate the r-th ($r \in \mathbb{N}$) order moments of (2.22.1), in terms of hypergeometric functions as follows:

$$G_{n,c}(e_r, x) = \begin{cases} \frac{nx \cdot r!}{(n-c)(n-2c)\cdots(n-rc)} \, _2F_1\left(\frac{n}{c} + 1, 1 - r; 2; -cx\right), & c \in \mathbb{N} \cup \{-1\}, \\ \frac{nx \cdot r!}{n^r} \, _1F_1(1 - r; 2; -nx), & c = 0. \end{cases}$$

We shall prove the theorem for three different cases separately:

Case $c = 0$ and $\phi_{n,c}(x) = e^{-nx}$: By the definition of $G_{n,c}$ for $c = 0$, the value of $G_{n,0}$ on the r-th power function (r-th moment), may be computed as

$$
G_{n,0}(e_r, x) = n \sum_{k=1}^{\infty} p_{n,k}(x, 0) \int_0^{\infty} p_{n,k-1}(t, 0) t^r dt
$$

$$
= n \sum_{k=1}^{\infty} e^{-nx} \frac{(nx)^k}{k!} \int_0^{\infty} e^{-nt} \frac{(nt)^{k-1}}{(k-1)!} t^r dt
$$

$$
= ne^{-nx} \sum_{k=1}^{\infty} \frac{(nx)^k}{k!(k-1)!} \int_0^{\infty} e^{-nt} n^{k-1} t^{k+r-1} dt
$$

$$
= ne^{-nx} \sum_{k=1}^{\infty} \frac{(nx)^k}{k!(k-1)!} \frac{\Gamma(k+r)}{n^{r+1}}
$$

$$
= ne^{-nx} \sum_{k=0}^{\infty} \frac{(nx)^{k+1}}{(k+1)!k!} \frac{\Gamma(k+r+1)}{n^{r+1}}.
$$

Using the fact that
$$
\Gamma(r+k+1) = \Gamma(r+1)(r+1)_k ,
$$

we have

$$
G_{n,0}(e_r, x) = \frac{n^2 x \Gamma(r+1)}{n^{r+1}} e^{-nx} \sum_{k=0}^{\infty} \frac{(nx)^k (r+1)_k}{(2)_k . k!}
$$

$$
= \frac{nx(r)!}{n^r} e^{-nx} \, _1F_1(r+1; 2; nx).
$$

Applying Kummer's transformation

$$
_1F_1(a; b; z) = e^z \, _1F_1(b - a; b; -z) .
$$

we get

$$
G_{n,0}(e_r, x) = \frac{nx \cdot r!}{n^r} \, _1F_1(1 - r; 2; -nx).
$$

Case $c \in \mathbb{N}$ and $\phi_{n,c}(x) = (1 + cx)^{-n/c}$: By definition of $G_{n,c}$; $c, n \in \mathbb{N}$, we can write

$$G_{n,c}(e_r, x) = n \sum_{k=1}^{\infty} p_{n,k}(x, c) \int_0^{\infty} p_{n+c,k-1}(t, c) t^r \, dt$$

$$= n \sum_{k=1}^{\infty} p_{n,k}(x, c) \int_0^{\infty} \frac{((n/c) + 1)_{k-1}}{(k-1)!} \frac{(ct)^{k-1}}{(1+ct)^{(n/c)+k}} t^r \, dt$$

$$= n \sum_{k=1}^{\infty} p_{n,k}(x, c) \frac{((n/c) + 1)_{k-1}}{(k-1)!} B(k+r, (n/c) - r) \left(\frac{1}{c}\right)^{r+1}$$

$$= n \left(\frac{1}{c}\right)^{r+1} \sum_{k=1}^{\infty} p_{n,k}(x, c) \frac{\Gamma(k + (n/c))}{\Gamma((n/c) + 1)(k-1)!} \frac{\Gamma(k+r)\Gamma((n/c) - r)}{\Gamma((n/c) + k)}$$

$$= n \Gamma((n/c) - r) \left(\frac{1}{c}\right)^{r+1} \sum_{k=1}^{\infty} p_{n,k}(x, c) \frac{\Gamma(k+r)}{\Gamma((n/c) + 1)(k-1)!}$$

$$= n \Gamma((n/c) - r) \left(\frac{1}{c}\right)^{r+1} \sum_{k=1}^{\infty} \frac{(n/c)_k}{k!} \frac{(cx)^k}{(1+cx)^{(n/c)+k}} \frac{\Gamma(k+r)}{\Gamma((n/c) + 1)(k-1)!}$$

$$= n \frac{\Gamma((n/c) - r)}{\Gamma((n/c) + 1)} \left(\frac{1}{c}\right)^{r+1} \sum_{k=1}^{\infty} \frac{(n/c)_k}{k!(k-1)!} \frac{(cx)^k}{(1+cx)^{(n/c)+k}} \Gamma(k+r)$$

$$= n \frac{\Gamma((n/c) - r)}{\Gamma((n/c) + 1)} \left(\frac{1}{c}\right)^{r+1} \sum_{k=0}^{\infty} \frac{(n/c)_{k+1}}{(k+1)!k!} \frac{(cx)^{k+1}}{(1+cx)^{(n/c)+k+1}} \Gamma(k+r+1)$$

$$= nx \frac{\Gamma((n/c) - r)\Gamma(r+1)}{\Gamma(n/c)c^r} (1+cx)^{-(n/c)-1} \sum_{k=0}^{\infty} \frac{((n/c) + 1)_k (r+1)_k}{(2)_k \cdot k!} \frac{(cx)^k}{(1+cx)^k}$$

$$= nx \frac{\Gamma((n/c) - r)\Gamma(r+1)}{\Gamma(n/c)c^r} (1+cx)^{-(n/c)-1} \; {}_2F_1\left(\frac{n}{c} + 1, r + 1; 2; \frac{cx}{1+cx}\right).$$

Applying Kummer's transformation

$$\quad {}_2F_1(a, b; c; z) = (1 - z)^{-a} \; {}_2F_1\left(a, c - b; c; \frac{z}{z-1}\right).$$

we get

$$G_{n,c}(e_r, x) = nx \frac{\Gamma((n/c) - r)\Gamma(r+1)}{\Gamma(n/c)} \left(\frac{1}{c}\right)^r \; {}_2F_1\left(\frac{n}{c} + 1, 1 - r; 2; -cx\right).$$

Case $c = -1$ and $\phi_{n,c}(x) = (1 - x)^n$: Using the identity

$$\binom{n}{k} = \frac{n!}{k!(n-k)!} = \frac{n(n-1)(n-2).....(n-k+1)}{k!} = \frac{(-1)^k(-n)_k}{k!},$$

for the case $c = -1$ the value of $G_{n,-1}$ on the r-th power function (r-th moment) is calculated below:

$$G_{n,-1}(e_r, x) = n \sum_{k=1}^{n} p_{n,k}(x, -1) \int_0^1 p_{n-1,k-1}(t, -1) t^r dt$$

$$= n \sum_{k=1}^{n} \binom{n}{k} x^k (1-x)^{n-k} \int_0^1 \binom{n-1}{k-1} t^{k+r-1} (1-t)^{n-k} dt$$

$$= n \sum_{k=1}^{n} \binom{n}{k} x^k (1-x)^{n-k} \frac{(n-1)!}{(k-1)!(n-k)!} \frac{(k+r-1)!(n-k)!}{(n+r)!}$$

$$= \frac{n!}{(n+r)!} (1-x)^n \sum_{k=1}^{n} \frac{(-1)^k(-n)_k}{k!} \left(\frac{x}{1-x}\right)^k \frac{(k+r-1)!}{(k-1)!}$$

$$= \frac{n!.nx}{(n+r)!} (1-x)^{n-1} \sum_{k=0}^{n} \frac{(-1)^k(-n+1)_k}{(2)_k} \left(\frac{x}{1-x}\right)^k \frac{(k+r)!}{k!}$$

$$= \frac{n!.nx.(r)!}{(n+r)!} (1-x)^{n-1} \sum_{k=0}^{n} \frac{(-n+1)_k(r+1)_k}{(2)_k} \left(\frac{x}{x-1}\right)^k \frac{1}{k!}$$

$$= \frac{n!.nx.(r)!}{(n+r)!} (1-x)^{n-1} {}_2F_1 \left(-n+1, r+1; 2; \frac{x}{x-1}\right),$$

where we have used $(m+k)! = (m+1)_k.m!$

Using Kummer's transformation

$$ {}_2F_1(a, b; c; z) = (1-z)^{-a} {}_2F_1 \left(a, c-b; c; \frac{z}{z-1}\right).$$

$$G_{n,-1}(e_r, x) = \frac{n!(nx) \cdot r!}{(n+r)!} {}_2F_1 (-n+1, 1-r; 2; x)$$

$$= \frac{(nx)r!}{(n+1)(n+2) \cdots (n+r)} {}_2F_1 (-n+1, 1-r; 2; x).$$

The operators defined by (2.22.1) reproduce only the constant functions for all $c \in \mathbb{N} \cup \{-1\} \cup \{0\}$, but for the case $c = 0$ these operators also reproduce the linear functions. Below, we mention some examples of modifications of Srivastava-Gupta operators, recently proposed by Gupta in [55], which preserve linear functions along with constant functions for all such c.

Example 2.1 For $c \in \mathbb{N} \cup \{0\}$ one can define

$$V_{n,c}(f, x) = n \sum_{k=1}^{\infty} p_{n-c,k}(x, c) \int_0^{\infty} p_{n+c,k-1}(t, c) f(t) dt$$
$$+ p_{n-c,0}(x, c) f(0),$$

where $p_{n,k}(x, c)$ is as defined in (2.22.1) above. The two cases mentioned above provide the Phillips operators and the genuine Baskakov-Durrmeyer type operators respectively. For $c = -1$ the operators take the form:

$$V_{n,-1}(f, x) = n \sum_{k=1}^{n} p_{n+1,k}(x, -1) \int_0^1 p_{n-1,k-1}(t, -1) f(t) dt$$
$$+ p_{n+1,0}(x, -1) f(0) + p_{n+1,n+1}(x, -1),$$

The r-th ($r \in \mathbb{N}$) order moments of $V_{n,c}$, in terms of hypergeometric functions, is as follows

$$V_{n,c}(e_r, x) = \begin{cases} x \dfrac{\Gamma((n/c) - r)\Gamma(r + 1)}{\Gamma((n/c) - 1).c^{r-1}} \, _2F_1\left(\frac{n}{c}, 1 - r; 2; -cx\right), & c \in \mathbb{N} \cup \{-1\}, \\[4mm] \frac{nx.r!}{n^r} \, _1F_1(1 - r; 2; -nx), & c = 0. \end{cases}$$

Remark 2.1 Few moments are given as:

$$V_{n,c}(e_0, x) = 1,$$
$$V_{n,c}(e_1, x) = x,$$
$$V_{n,c}(e_2, x) = \frac{x(2 + nx)}{(n - 2c)},$$
$$V_{n,c}(e_3, x) = \frac{x[n(n + c)x^2 + 6nx + 6]}{(n - 2c)(n - 3c)},$$
$$V_{n,c}(e_4, x) = \frac{x[n(n^2 + 2nc + 2c^2)x^3 + 12n(n + c)x^2 + 36nx + 24]}{(n - 2c)(n - 3c)(n - 4c)}.$$

Example 2.2 For $c \in \mathbb{N} \cup \{0\}$ the genuine operators are defined by

$$S_{n,c}(f, x) = (n + c) \sum_{k=1}^{\infty} p_{n,k}(x, c) \int_0^{\infty} p_{n+2c,k-1}(t, c) f(t) dt$$
$$+ p_{n,0}(x, c) f(0),$$

For $c = -1$, one may consider

$$S_{n,-1}(f, x) = (n-1) \sum_{k=1}^{n-1} p_{n,k}(x, -1) \int_0^1 p_{n-2,k-1}(t, -1) f(t) dt$$
$$+ p_{n,0}(x, -1) f(0) + p_{n,n}(x, -1),$$

where $p_{n,k}(x, c)$ is as defined in (2.22.1) above. This form also preserves linear function.

After simple calculations we obtain the value of $S_{n,c}$ on the r-th power functions (r-th order moments), for $r \in \mathbb{N}$:

$$S_{n,c}(e_r, x) = \begin{cases} x \frac{\Gamma((n/c)-r+1)\Gamma(r+1)}{\Gamma(n/c).c^{r-1}} \, {}_2F_1\left(\frac{n}{c}+1, 1-r; 2; -cx\right), & c \in \mathbb{N} \cup \{-1\}, \\ \frac{nx.r!}{n^r} \, {}_1F_1(1-r; 2; -nx), & c = 0. \end{cases}$$

Actually the operators $S_{n,c}$ are similar to the operators $\overline{V}_{n,\alpha}$ defined in (2.21.1) for $\alpha = c$. These are represented in different forms. Gupta-Sinha in [72] discussed only the case $\alpha > 0$, which is further extended for different values as mentioned in Example 2.2.

Remark 2.2 For a special value $c = 1$, these forms of operators can be written as a composition of Baskakov operators (1.2.1) and the Stancu-Beta operators (2.5.1), i.e.

$$S_{n,1} = V_n \circ \overline{A}_n.$$

Also, for $c = -1$, these operators constitute a composition of Bernstein polynomials (1.1.1) and Beta operators of the first kind (2.6.1), i.e.

$$S_{n,-1} = B_n \circ \overline{C}_n.$$

Example 2.3 Yet another example of the genuine operators is defined for $c \in \mathbb{N} \cup \{0\}$ as

$$R_{n,c}(f, x) = (n+3c) \sum_{k=1}^{\infty} p_{n+2c,k}(x, c) \int_0^{\infty} p_{n+4c,k-1}(t, c) f(t) dt$$
$$+ p_{n+2c,0}(x, c) f(0),$$

For $c = -1$, one may consider

$$R_{n,-1}(f, x) = (n-3) \sum_{k=1}^{n-3} p_{n-2,k}(x, -1) \int_0^1 p_{n-4,k-1}(t, -1) f(t) dt$$
$$+ p_{n-2,0}(x, -1) f(0) + p_{n-2,n-2}(x, -1),$$

where $p_{n,k}(x, c)$ is as defined in (2.22.1) above. This form also preserves linear function.

The r-th ($r \in \mathbb{N}$) order moments of $R_{n,c}$ are given by

$$R_{n,c}(e_r, x) = \begin{cases} x \frac{\Gamma((n/c)-r+3)\Gamma(r+1)}{\Gamma((n/c)+2).c^{r-1}} \, {}_2F_1\left(\frac{n}{c}+3, 1-r; 2; -cx\right), & c \in \mathbb{N} \cup \{-1\}, \\ \frac{nx.r!}{n^r} \, {}_1F_1(1-r; 2; -nx), & c = 0. \end{cases}$$

Example 2.4 If $c \in \mathbb{N} \cup \{0\}$, then the genuine operators are defined by

$$U_{n,c}(f, x) = (n + 2c) \sum_{k=1}^{\infty} p_{n+c,k}(x, c) \int_0^{\infty} p_{n+3c,k-1}(t, c) f(t)dt$$
$$+ p_{n+c,0}(x, c) f(0),$$

For $c = -1$, one may consider

$$U_{n,-1}(f, x) = (n - 2) \sum_{k=1}^{n-3} p_{n-1,k}(x, -1) \int_0^{1} p_{n-3,k-1}(t, -1) f(t)dt$$
$$+ p_{n-1,0}(x, -1) f(0) + p_{n-1,n-1}(x, -1),$$

where $p_{n,k}(x, c)$ is as defined in (2.22.1) above. This form also preserves linear function.

The r-th ($r \in \mathbb{N}$) order moments of $U_{n,c}$ are given by

$$U_{n,c}(e_r, x) = \begin{cases} x \frac{\Gamma((n/c)-r+2)\Gamma(r+1)}{\Gamma((n/c)+1).c^{r-1}} \, {}_2F_1\left(\frac{n}{c}+2, 1-r; 2; -cx\right), & c \in \mathbb{N} \cup \{-1\}, \\ \frac{(nx)r!}{n^r} \, {}_1F_1(1-r; 2; -nx), & c = 0. \end{cases}$$

The operators presented in the above Examples 2.1–2.4, may be called genuine operators and can provide better approximation over the Srivastava-Gupta operators (2.22.1). From the above examples it is observed that the suffix n in the basis function $p_{n,k}(x, c)$ has a difference of $2c$ under summation and integral sign in order to obtain the genuine operators. Very recently Gupta and Srivastava in [77] provided a general sequence of genuine operators, which for an integer m and $c \in \mathbb{N} \cup \{0\}$ is defined as

$$\overline{G}_{n,c}(f, x) = [n + (m + 1)c] \sum_{k=1}^{\infty} p_{n+mc,k}(x, c) \tag{2.22.2}$$

$$\int_0^{\infty} p_{n+(m+2)c,k-1}(t, c) f(t)dt + p_{n+mc,0}(x, c) f(0),$$

where $p_{n,k}(x, c)$ is as defined in (2.22.1) above. For $c = -1$ the operators take the form:

$$\overline{G}_{n,-1}(f, x) = (n - m - 1) \sum_{k=1}^{n-m-1} p_{n-m,k}(x, -1) \int_0^1 p_{n-m-2,k-1}(t, -1) f(t) dt$$

$$+ p_{n-m,0}(x, -1) f(0) + p_{n-m,n-m}(x, -1),$$

The r-th ($r \in \mathbb{N}$) order moments of (2.22.2), satisfy

$$\overline{G}_{n,c}(e_r, x) = \begin{cases} x \frac{\Gamma((n/c)-r+m+1)\Gamma(r+1)}{\Gamma((n/c)+m).c^{r-1}} \, _2F_1\left(\frac{n}{c} + m + 1, 1 - r; 2; -cx\right), & c \in \mathbb{N} \cup \{-1\}, \\ \frac{nx.r!}{n^r} \, _1F_1(1 - r; 2; -nx), & c = 0. \end{cases}$$

Recently some approximation results (which include direct results and rate of convergence) for the Gupta-Srivastava operators $\overline{G}_{n,c}$ have been discussed by Pratap and Deo [122].

2.23 Genuine Lupaş-Beta Operators

In [70] Gupta-Rassias-Yadav considered the following form of hybrid operators for $x \geq 0$, which preserve constant as well linear functions

$$\overline{Y}_n(f, x) = \sum_{k=1}^{\infty} l_{n,k}(x) \int_0^{\infty} b_{n,k-1}(t) f(t) dt + 2^{-nx} f(0), \qquad (2.23.1)$$

where

$$l_{n,k}(x) = 2^{-nx} \frac{(nx)_k}{k! \cdot 2^k}, \qquad b_{n,k-1}(t) = \frac{1}{B(k, n+1)} \frac{t^{k-1}}{(1 + t)^{k+n+1}}$$

and $B(m, n)$ being the Beta function.

It was observed in [70] that

$$\overline{Y}_n(e_r, x) = \frac{r!(n - r)!.(nx)}{n!} \cdot _2F_1(nx + 1, 1 - r; 2; -1).$$

Recently Gupta-Rassias-Pandey in [71] obtained the moments of the operators (2.23.1) using an alternate approach. They used factorial polynomials defined by

$$k^{(m)} = k(k - 1)(k - 2) \cdots (k - m + 1)$$

and the elementary hypergeometric functions $_1F_0(a; -; x)$ and obtained few moments, as follows

$$\overline{Y}_n(e_0, x) = 1, \overline{Y}_n(e_1, x) = x, \overline{Y}_n(e_2, x) = \frac{nx^2 + 3x}{n - 1}$$

$$\overline{Y}_n(e_3, x) = \frac{n^2x^3 + 9nx^2 + 14x}{(n - 1)(n - 2)}, \overline{Y}_n(e_4, x) = \frac{n^3x^4 + 18n^2x^3 + 83nx^2 + 90x}{(n - 1)(n - 2)(n - 3)},$$

$$\overline{Y}_n(e_5, x) = \frac{n^4x^5 + 30n^3x^4 + 275n^2x^3 + 870nx^2 + 744x}{(n - 1)(n - 2)(n - 3)(n - 4)},$$

and

$$\overline{Y}_n(e_6, x) = \frac{n^5x^6 + 45n^4x^5 + 685n^3x^4 + 4275n^2x^3 + 10474nx^2 + 7560x}{(n - 1)(n - 2)(n - 3)(n - 4)(n - 5)}.$$

Chapter 3
Approximation Properties of Certain Operators

The present chapter deals with the approximation properties of certain integral type operators. The direct results are presented for the hybrid Baskakov-Szász-Mirakyan operators, preserving some exponential type operators. The chapter consists of the study of direct results for the Post-Widder operators preserving the general test function e_r, $r \in \mathbb{N}$. The linear combinations have been discussed for certain operators, which speed up the convergence. Also, some other direct results have been presented for the Kantorovich type operators, which have been recently discussed.

3.1 Certain Baskakov-Szász-Mirakyan Operators

Baskakov-Szász-Mirakyan operators introduced in [12] are defined as follows:

$$B_n(f; x) = n \sum_{k=1}^{\infty} v_{n,k}(x) \int_0^{\infty} s_{n,k-1}(t) f(t) dt + b_{n,0}(x) f(0), \qquad (3.1.1)$$

where

$$v_{n,k}(x) = \frac{(n)_k}{k!} \frac{x^k}{(1+x)^{n+k}}, \quad s_{n,k}(t) = e^{-nt} \frac{(nt)^k}{(k)!},$$

and

$$(n)_0 = 1, \quad (n)_k = n(n+1) \cdots (n+k-1) \text{ for } k \geq 1.$$

Some approximation properties of these operators and their different versions have been discussed in [54, 60]. Using the technique proposed in [7, 62, 79], very recently Gupta and Acu [59] modified the operators defined in (3.1.1) as follows:

© The Author(s), under exclusive license to Springer Nature Switzerland AG 2019
V. Gupta and M. T. Rassias, *Moments of Linear Positive
Operators and Approximation*, SpringerBriefs in Mathematics,
https://doi.org/10.1007/978-3-030-19455-0_3

$$A_n(f, x) = n \sum_{k=1}^{\infty} \frac{(n)_k}{k!} (1 + a_n(x))^{-n} \left(\frac{b_n(x)}{1 + b_n(x)} \right)^k$$

$$\int_0^{\infty} s_{n,k-1}(t) \, f(t) \, dt + (1 + a_n(x))^{-n} f(0). \qquad (3.1.2)$$

In the case when $a_n(x) = b_n(x) = x$, we derive the operators due to Agrawal and Mohammad [12]. Suppose that operators (3.1.2) reproduce e^{ax} and e^{bx}. After simple computation, we obtain

$$n \int_0^{\infty} e^{-nt} \frac{(nt)^{k-1}}{(k-1)!} e^{At} \, dt = \left(\frac{n}{n-A} \right)^k, \quad \text{for } n > A. \qquad (3.1.3)$$

Thus, using (3.1.3) and the well known binomial series

$$\sum_{k=0}^{\infty} \frac{(a)_k}{k!} z^k = (1 - z)^{-a}, \, |z| < 1,$$

we have

$$A_n(e^{at}, x) = e^{ax} = \left[\frac{1 + b_n(x)}{1 + a_n(x)} \right]^n \left[\frac{(n-a)}{n - a(1 + b_n(x))} \right]^n.$$

Similarly

$$A_n(e^{bt}, x) = e^{bx} = \left[\frac{1 + b_n(x)}{1 + a_n(x)} \right]^n \left[\frac{(n-b)}{n - b(1 + b_n(x))} \right]^n.$$

These imply

$$b_n(x) = \frac{(n^2 - na - nb + ab)(e^{(a-b)x/n} - 1)}{a(n-b)e^{(a-b)x/n} - b(n-a)}$$

and

$$a_n(x) = \frac{(n-b)e^{-bx/n} - (n-a)e^{-ax/n} - (a-b)}{a-b}.$$

The present section is devoted to the study for $a = 0$, $b = -1$ and $a = 0$, $b = -2$. In these two cases the operator (3.1.2) is positive and preserves the exponential functions e^{ax} and e^{bx}, respectively.

Here $C^*[0, \infty)$ denotes the class of real-valued continuous functions $f(x)$, possessing finite limit for x sufficiently large and equipped with the uniform norm. Holhoş [86] considered the following quantitative estimate for a sequence L_n of positive linear operators.

Theorem A ([86]) *For a sequence of positive linear operators $L_n : C^*[0, \infty) \to$ $C^*[0, \infty)$, if for $i = 0, 1, 2$ we denote the norms $||L_n(e^{-it}) - e^{-ix}|_{[0,\infty)}$ as α_n, β_n and γ_n respectively, and $\alpha_n, \beta_n, \gamma_n$ tend to 0 as $n \to \infty$, then*

$$||L_n f - f||_{[0,\infty)} \le \alpha_n ||f||_{[0,\infty)} + (2 + \alpha_n)\omega^*(f, \sqrt{\alpha_n + 2\beta_n + \gamma_n}),$$

where the modulus of continuity is given by

$$\omega^*(f, \delta) := \sup_{\substack{|e^{-x} - e^{-t}| \le \delta \\ x,t > 0}} |f(t) - f(x)|.$$

Corresponding to Theorem A, the operators A_n satisfy the following:

Theorem 3.1 ([59]) *For $f \in C^*[0, \infty)$, the following hold true:*

(i) If $a = 0, b = -1$, then

$$||A_n f - f||_{[0,\infty)} \le 2\omega^*(f, \sqrt{\gamma_n}), \tag{3.1.4}$$

where

$$\gamma_n = ||A_n(e^{-2t}) - \varphi_2||_{[0,\infty)} \to 0, n \to \infty.$$

(ii) If $a = 0, b = -2$, then

$$||A_n f - f||_{[0,\infty)} \le 2\omega^*(f, \sqrt{2\beta_n}), \tag{3.1.5}$$

where

$$\beta_n = ||A_n(e^{-t}) - \varphi_1||_{[0,\infty)} \to 0, n \to \infty.$$

Remark 3.1 For $f \in C^*[0, \infty)$, the original Baskakov-Szász-Mirakyan operators verify:

$$||B_n f - f||_{[0,\infty)} \le 2\omega^* \left(f, \sqrt{2\tilde{\beta}_n + \tilde{\gamma}_n} \right), \tag{3.1.6}$$

where

$$\tilde{\beta}_n = \sup_{x \in [0,\infty)} |B_n(e^{-t}) - e^{-x}| \le \frac{1}{n} \left(2e^{-2} + e^{-1} \right) + \mathcal{O}\left(\frac{1}{n^2}\right) \le \mathcal{O}(n^{-1}),$$

$$\tilde{\gamma}_n = \sup_{x \in [0,\infty)} |B_n(e^{-2t}) - e^{-2x}| \le \frac{2}{n} \left(e^{-2} + e^{-1} \right) + \mathcal{O}\left(\frac{1}{n^2}\right) \le \mathcal{O}(n^{-1}).$$

Table 3.1 Error of approximation for A_n and B_n

x	$|A_n(f,x) - f(x)|$ for $a = 0, b = -1$	$|A_n(f,x) - f(x)|$ for $a = 0, b = -2$	$|B_n(f,x) - f(x)|$
1.6	0.000895697810	0.006904613420	0.009293893320
1.8	0.002968634830	0.004227034830	0.011138359620
2.0	0.003929060620	0.002376955660	0.011484189280
2.2	0.004131642890	0.001187695670	0.010866295170
2.4	0.003881831328	0.000475399700	0.009721775838
2.6	0.003405286475	0.000082338386	0.008363234064
2.8	0.002851083529	0.000110673223	0.006991228846
3.0	0.002307651666	0.000186425999	0.005720090824
3.2	0.001820641481	0.000198739671	0.004604289701
3.4	0.001408169148	0.000180545612	0.003660211511
3.6	0.001072178275	0.000150745278	0.002882178393
3.8	0.000806167696	0.000119356635	0.002253297893
4.0	0.000600056152	0.000091081562	0.001752296622

Since $\gamma_n \leq \tilde{\gamma}_n$ and $\beta_n \leq \tilde{\beta}_n$, from the relations (3.1.4), (3.1.5), (3.1.6) it follows that the modified Baskakov-Szász-Mirakyan operators A_n give better approximation than the usual Baskakov-Szász-Mirakyan operators B_n (Table 3.1).

The next result, established in [59] is a quantitative form of Voronovskaya's formula :

Theorem 3.2 ([59]) *Let f, $f'' \in C^*[0, \infty)$, then for any $x \in [0, \infty)$, we have*

$$\left| n[A_n(f,x) - f(x)] + \left(ax + bx + \frac{ax^2}{2} + \frac{bx^2}{2} \right) f'(x) - \left(x + \frac{x^2}{2} \right) f''(x) \right|$$
$$\leq |p_n(x)| \cdot |f'(x)| + |q_n(x)| \cdot |f''(x)| + 2\left[2q_n(x) + (x^2 + 2x) + r_n(x) \right] \omega^*(f'', n^{-1/2}),$$

where

$$p_n(x) = nT_{n,1}^A(x) + \left(ax + bx + \frac{ax^2}{2} + \frac{bx^2}{2} \right)$$

$$q_n(x) = \frac{1}{2}[nT_{n,2}^A(x) - (x^2 + 2x)]$$

$$r_n(x) = n^2[A_n((e^{-x} - e^{-t})^4, x) \cdot T_{n,4}^A(x)]^{1/2}$$

and $T_{n,m}^A(x) = A_n((t - x)^m, x)$.

Remark 3.2 By direct calculations the following results can be obtained

(i) $\lim\limits_{n\to\infty} n^2 T_{n,4}^A(x) = 3x^2(x+2)^2$,

(ii) $\lim\limits_{n\to\infty} n^2 A_n \left(\left(e^{-x} - e^{-t}\right)^4, x \right) = 3x^2(x+2)^2 e^{-4x}$.

As a consequence of the Theorem 3.2 we derive the following result:

Corollary 3.1 ([59]) *Let* $f, f'' \in C^*[0, \infty)$, *then for* $x \in [0, \infty)$ *we have*

$$\lim_{n\to\infty} n \left[A_n(f, x) - f(x) \right] = - \left(ax + bx + \frac{ax^2}{2} + \frac{bx^2}{2} \right) f'(x) + \left(x + \frac{x^2}{2} \right) f''(x).$$

3.2 Post-widder Operators

The Post-Widder operators for $n \in \mathbb{N}$ and $x > 0$ considered by Widder [140] are defined by

$$P_n(f, x) := \frac{1}{n!} \left(\frac{n}{x}\right)^{n+1} \int_0^\infty t^n e^{-\frac{nt}{x}} f(t) \, dt.$$

These operators preserve constant functions only. The q analogue of these operators was recently studied by Aydin et al. [17]. Rathore and Singh [124] considered a parameter p, while defining Post-Widder operators. Rempulska and Skorupka in [125] further modified the Post-Widder operators of the form considered by May [103] in order to preserve the test function e_2, where $e_r(x) = x^r$. It was observed in [125] that the modified form provides better approximation results over the form of [103], but in that case the modified form loses the preservation of the test function e_1. It may be observed that only two preservations can be made at a time: either constant and e_1 or constant and any other order. Recently Gupta–Tachev in [81] considered the modification of Post-Widder operators which preserve the constant and $e_r, r \in \mathbb{N}$.

Following [125], the r-th order moments

$$\mu_r^{P_n}(x) = P_n(e_r, x), \text{ where } e_r(t) = t^r, r \in \mathbb{N} \cup \{0\}$$

are given by

$$\mu_r^{P_n}(x) = \frac{(n+1)_r x^r}{n^r}, \tag{3.2.1}$$

where $(n)_r = n(n+1)(n+2) \cdots (n+r-1)$ is the the rising factorial, with $(n)_0 = 1$.

If the central moments are denoted by

$$T_m^{P_n}(x) = P_n((t - x)^m, x),$$

then

$$T_1^{P_n}(x) = \frac{x}{n},$$

$$T_2^{P_n}(x) = \frac{(n + 2)x^2}{n^2}.$$

Following the results given in [32, Chap. 9], one can deduce that for every continuous and bounded function f on $(0, \infty)$, it holds

$$|P_n(f, x) - f(x)| \le \omega\left(f, \frac{\sqrt{n + 2}}{n}x\right). \qquad (3.2.2)$$

Let us consider that the Post-Widder operators preserve the test function x^r, $r \in \mathbb{N}$. Then we start with the following form

$$\widetilde{P}_{n,r}(f, x) = \frac{1}{n!}\left(\frac{n}{a_{n,r}(x)}\right)^{n+1} \int\limits_0^\infty t^n e^{-\frac{nt}{a_{n,r}(x)}} f(t)dt.$$

Hence

$$\widetilde{P}_{n,r}(e_r, x) = x^r = \frac{1}{n!}\left(\frac{n}{a_{n,r}(x)}\right)^{n+1} \int\limits_0^\infty t^{n+r} e^{-\frac{nt}{a_{n,r}(x)}} dt$$

$$= \frac{(n + r)!}{n!}\left(\frac{a_{n,r}(x)}{n}\right)^r = (n + 1)_r \left(\frac{a_{n,r}(x)}{n}\right)^r,$$

implying that

$$a_{n,r}(x) = \frac{nx}{((n + 1)_r)^{1/r}}.$$

Thus the modified Post-Widder operators $\widetilde{P}_{n,r}$, $r \in \mathbb{N}$ discussed in [81] assume the following form

$$\widetilde{P}_{n,r}(f, x) := \frac{1}{n!}\left[\frac{((n + 1)_r)^{1/r}}{x}\right]^n \int_0^\infty t^n e^{-\frac{t}{x}((n+1)_r)^{1/r}} f(t)\, dt,$$

which preserve the function x^r and the constant function.

Following (3.2.1), the r-th order moments are given by

$$\tilde{P}_{n,r}(e_m, x) := \frac{(n+1)_m \cdot (a_{n,r}(x))^m}{n^m} = \frac{(n+1)_m}{((n+1)_r)^{m/r}} x^m.$$

Suppose the operators preserve the test functions e_1, e_2, e_3, e_4, then (see [81]), we respectively have for every continuous and bounded function f on $(0, \infty)$, the following:

$$|P_{n,1}(f, x) - f(x)| \leq \omega\left(f, \frac{x}{\sqrt{n+1}}\right).$$

$$|P_{n,2}(f, x) - f(x)| \leq \omega\left(f, \sqrt{2}\sqrt{\left(1 - \sqrt{\frac{n+1}{n+2}}\right)x}\right).$$

$$|P_{n,3}(f, x) - f(x)|$$
$$\leq \omega\left(f, \sqrt{\left(\frac{[(n+1)(n+2)]^{1/3}}{(n+3)^{2/3}} - 2\frac{(n+1)^{2/3}}{[(n+2)(n+3)]^{1/3}} + 1\right).x}\right).$$

$$|P_{n,4}(f, x) - f(x)|$$
$$\leq \omega\left(f, \sqrt{\left(\frac{[(n+1)(n+2)]^{1/2}}{[(n+3)(n+4)]^{1/2}} - 2\frac{(n+1)^{3/4}}{[(n+2)(n+3)(n+4)]^{1/4}} + 1\right).x}\right).$$

If we compare the above results, with the estimate (3.2.2), we find that the error becomes smaller and monotonically decreasing for $n \in \mathbb{N}, x \in (0, \infty)$ until the preservation of the test function e_3, as the following holds true for second order moments:

$$\frac{\sqrt{n+2}}{n} \geq \frac{1}{\sqrt{n+1}} \geq \sqrt{2}\sqrt{\left(1 - \sqrt{\frac{n+1}{n+2}}\right)}$$

$$\geq \sqrt{\left(\frac{[(n+1)(n+2)]^{1/3}}{(n+3)^{2/3}} - 2\frac{(n+1)^{2/3}}{[(n+2)(n+3)]^{1/3}} + 1\right)}.$$

But, for higher order preservation of test functions i.e. preservation of e_4, one cannot obtain better approximation, which is also shown in the table above, although we have convergence in all cases for n sufficiently large.

Approximation for different test function's preservation:

n	$\sqrt{T_2^{\tilde{P}_{n,1}}(x)}$	$\sqrt{T_2^{\tilde{P}_{n,2}}(x)}$	$\sqrt{T_2^{\tilde{P}_{n,3}}(x)}$	$\sqrt{T_2^{\tilde{P}_{n,4}}(x)}$
1	0.7071067814x	0.6058108929x	0.5782752253x	0.5823850946x
2	0.5773502690x	0.5176380898x	0.5003545374x	0.5049814095x
3	0.5000000000x	0.4595058411x	0.4473814474x	0.4517310256x
10	0.3015113446x	0.2917975051x	0.2886837212x	0.2907399443x
20	0.2182178902x	0.2144368392x	0.2132017781x	0.2141867802x
100	0.09950371903x	0.09913661786x	0.09901475698x	0.09913079743x
1000	0.03160697706x	0.03159514203x	0.03159119181x	0.03159513095x

Let $C_B[0, \infty)$ denote the space of bounded and continuous functions on $[0, \infty)$ endowed with the norm

$$\|f\| = \sup\{|f(x)| : x \in [0, \infty)\}.$$

Additionally, let us consider the following K-functional:

$$K_2(f, \delta) = \inf_{h \in C_B^2[0,\infty)} \{\|f - h\| + \delta\|h''\|\},$$

where $\delta > 0$ and

$$C_B^2[0, \infty) = \{h \in C_B[0, \infty) : h', h'' \in C_B[0, \infty)\}.$$

Theorem 3.3 ([81]) *Let* $f \in C_B[0, \infty)$. *Then*

$$\left|\tilde{P}_{n,r}(f, x) - f(x)\right| \leq C\omega_2\left(f, \sqrt{\delta_{n,r}}\right) + \omega\left(f, \left|\frac{(n+1)}{((n+1)_r)^{1/r}} - 1\right|x\right)$$

where C is a positive constant and

$$\delta_{n,r} = \left[\frac{(n+1)(2n+3)}{((n+1)_r)^{2/r}} - \frac{4(n+1)}{((n+1)_r)^{1/r}} + 2\right]x^2.$$

Corollary 3.2 ([81]) *Let* $f \in C_B[0, \infty)$. *Then*

$$\left|\tilde{P}_{n,1}(f, x) - f(x)\right| \leq C\omega_2\left(f, \frac{x}{\sqrt{(n+1)}}\right)$$

where C is some positive constant.

Let us consider

$$B_2[0, \infty) := \{f : f \in \mathbb{R}^+ \text{ and } |f(x)| \leq C(f)(1 + x^2), C(f) > 0\}$$

and
$$C_2 [0, \infty) = C [0, \infty) \cap B_2 [0, \infty) .$$

We set $C_2^k [0, \infty)$ to stand for the subspace of all continuous functions $f \in B_2 [0, \infty)$ for which
$$\lim_{x \to \infty} \frac{f(x)}{1 + x^2} < \infty.$$

The weighted modulus of continuity $\Omega (f, \delta)$ defined on the infinite interval $\mathbb{R}^+$ (see [8]) is given by

$$\Omega (f, \delta) = \sup_{|h| < \delta, \, x \in \mathbb{R}^+} \frac{|f (x + h) - f (x)|}{(1 + h^2 + x^2 + h^2 x^2)} \text{ for each } f \in C_2 [0, \infty) .$$

We now estimate the following quantitative Voronovskaja-type asymptotic formula:

Theorem 3.4 ([81]) *Let* $f'' \in C_2^k [0, \infty)$, *and* $r \in \mathbb{N}$, *then for* $x > 0$, *we have*

$$\left| \widetilde{P}_{n,r}(f, x) - f(x) - \left[\frac{(n+1)}{((n+1)_r)^{1/r}} - 1 \right] x f'(x) \right.$$
$$- \left[\frac{(n+1)_2}{((n+1)_r)^{2/r}} - 2 \frac{(n+1)}{((n+1)_r)^{1/r}} + 1 \right] \frac{x^2}{2} f''(x) \Bigg|$$
$$\leq C \left(1 + x^2 \right) \Omega \left(f'', n^{-1/2} \right) \left[\left(\frac{(n+1)_2}{((n+1)_r)^{2/r}} - 2 \frac{(n+1)}{((n+1)_r)^{1/r}} + 1 \right) x^2 \right.$$
$$+ n^2 x^6 \left(\frac{(n+1)_6}{((n+1)_r)^{6/r}} - 6 \frac{(n+1)_5}{((n+1)_r)^{5/r}} + 15 \frac{(n+1)_4}{((n+1)_r)^{4/r}} \right.$$
$$- 20 \frac{(n+1)_3}{((n+1)_r)^{3/r}} + 15 \frac{(n+1)_2}{((n+1)_r)^{2/r}} - 6 \frac{(n+1)}{((n+1)_r)^{1/r}} + 1 \Bigg) \Bigg],$$

where C *is some absolute constant.*

3.3 Combinations of Genuine Baskakov-Durrmeyer Operators

Recently in [83] the following linear combinations of the operators (2.21.1) have been studied:

$$\overline{V}_{n,r,\alpha} = \sum_{i=0}^{r} \alpha_i(n) \cdot \overline{V}_{n_i,\alpha}, \tag{3.3.1}$$

where n_i, $i = 0, 1, \ldots, r$, are distinct positive numbers. Here the goal is to determine $\alpha_i(n)$ in (3.3.1), such that

$$\overline{V}_{n,r,\alpha} p = p \text{ for all } p \in \mathbf{P}_{r+1} \, .$$

The operators $\overline{V}_{n,\alpha}$ preserve only the linear functions. Here the requirement is that each polynomial of degree at most $r + 1$ should be reproduced, which leads to a linear system of equations, i.e.

$$\overline{V}_{n,r,\alpha}(e_k, x) = x^k, \ 0 \le k \le r + 1.$$

Therefore the system

$$\alpha_0 + \alpha_1 + \cdots + \alpha_r = 1$$

$$\sum_{i=0}^{r} \alpha_i \cdot \frac{\Gamma(\frac{n_i}{\alpha}+j)\Gamma(\frac{n_i}{\alpha}-j+1)}{\Gamma(\frac{n_i}{\alpha}+1)\Gamma(\frac{n_i}{\alpha})} = 1, \ 2 \le j \le r + 1,$$

has the unique solution

$$\alpha_i = \alpha^r \left(\frac{n_i}{\alpha} - r\right)_r \cdot \prod_{\substack{j=0 \\ j \ne i}}^{r} \frac{1}{(n_i - n_j)}, \ 0 \le i \le r.$$

To obtain a direct estimate for $\overline{V}_{n,r,\alpha}$, one needs the following additional assumptions:

$$an = n_0 < n_1 < \cdots < n_r \le A \cdot n, (A = A(r)), \qquad (3.3.2)$$

$$\sum_{i=0}^{r} |\alpha_i(n)| \le C. \qquad (3.3.3)$$

The first of these conditions (3.3.2) guarantees that

$$\left(\overline{V}_{n,r,\alpha}|\psi_x^r|\right)(x) = O\left(n^{-\frac{r}{2}}\right), \ n \to \infty, \qquad (3.3.4)$$

where

$$\psi_x^r = (e_1 - xe_0)^r \, .$$

This follows from the fact that

$$\overline{V}_{n,\alpha}((e_1 - xe_0)^m, x) = O\left(n^{-[(m+1)/2]}\right).$$

The second condition (3.3.3) is that the sum of the absolute values of the coefficients should be bounded and independent of n. The following direct estimates have been discussed in [83]. Let $C_B[0, \infty)$ be the space of all real valued continuous and bounded functions f defined on $[0, \infty)$. The classical Peetre's K_r-functional for $f \in C_B[0, \infty)$ is defined by

$$K_r(f, \delta^r) = \inf\{\|f - g\| + \delta^r \cdot \|g^{(r)}\| : g \in W_\infty^r\}, \ \delta > 0, \qquad (3.3.5)$$

where

$$W_\infty^r = \{g \in C_B[0, \infty), g^{(r)} \in C_B[0, \infty)\}.$$

It is evident, following [29], that there exists a positive constant C such that

$$K_r(f, \delta^r) \le C\omega_r(f, \delta). \qquad (3.3.6)$$

Theorem 3.5 *Let* $f \in C_B[0, \infty)$. *Then for every* $x \in [0, \infty)$ *and for* $C > 0, n > r$ *we have*

$$|(\overline{V}_{n,r,\alpha}f)(x) - f(x)| \le C \cdot \omega_{r+2}\left(f, \frac{1}{\sqrt{n}}\right).$$

Proof Let $g \in W_\infty^{r+2}$. By the Taylor expansion of g we get

$$\overline{V}_{n,r,\alpha}(g, x) - g(x) = \overline{V}_{n,r,\alpha}\left(\frac{(t-x)^{r+2}}{(r+2)!} \cdot g^{(r+2)}(\xi_{t,x}); x\right)$$

Therefore

$$|\overline{V}_{n,r,\alpha}(g, x) - g(x)| \le \sum_{i=0}^{r} |\alpha_i| \cdot \overline{V}_{n_i}\left(\frac{|t-x|^{r+2}}{(r+2)!}; x\right) \cdot \|g^{(r+2)}\|_{C_B[0,\infty)}$$

From (3.3.3) and (3.3.4) it follows that

$$|\overline{V}_{n,r,\alpha}(g, x) - g(x)| \le C(r) \cdot n^{-\frac{r+2}{2}} \cdot \|g^{(r+2)}\|.$$

Consequently

$$|\overline{V}_{n,r,\alpha}(f, x) - f(x)| \le |\overline{V}_{n,r,\alpha}(f - g, x) - (f - g)(x)| + |\overline{V}_{n,r,\alpha}(g, x) - g(x)|$$

$$\le 2\|f - g\| + C(r) \cdot n^{-\frac{r+2}{2}} \cdot \|g^{(r+2)}\|.$$

Taking the infimum on the right hand side over all $g \in W_\infty^{r+2}$ and using (3.3.5), (3.3.6) we get the required result. ∎

Corollary 3.3 *If* $f^{(r+2)} \in C_B[0, \infty)$ *then*

$$|(\overline{V}_{n,r,\alpha}f)(x) - f(x)| \le C \cdot \left(\frac{1}{\sqrt{n}}\right)^{r+2} \cdot \|f^{(r+2)}\|_{C_B[0,\infty)}.$$

Theorem 3.6 *Let* $f, f', \ldots, f^{(r+2)} \in C_B[0, \infty)$. *Then, if* $r = 2k + 1, \ k = 0, 1, 2,$ *$\ldots$ for* $x \in [0, \infty)$ *it follows that*

$$\lim_{n \to \infty} n^{k+2} \cdot [\overline{V}_{n,2k+1,\alpha}(f, x) - f(x)] = Q_{2k+3}(x) \cdot f^{(2k+3)}(x),$$

where

$$Q_{2k+3}(x) = \lim_{n \to \infty} n^{k+2} \overline{V}_{n,2k+1,\alpha}(\psi_x^{2k+3}(t), x).$$

Proof By the Taylor expansion of f, we obtain

$$f(t) = f(x) + \sum_{i=1}^{2k+3} \frac{(t-x)^i}{i!} \cdot f^{(i)}(x) + \frac{(t-x)^{2k+3}}{(2k+3)!} \cdot R(t, x), \qquad (3.3.7)$$

where $R(t, x)$ is a bounded function for all $t, x \in [0, \infty)$ and

$$\lim_{t \to x} R(t, x) = 0.$$

We apply $\overline{V}_{n,2k+1,\alpha}$ to both sides of (3.3.7) to obtain

$$\overline{V}_{n,2k+1,\alpha}(f, x) - f(x) = \frac{f^{(2k+3)}(x)}{(2k+3)!} \cdot \overline{V}_{n,2k+1,\alpha}(\psi_x^{2k+3}, x) + I, \qquad (3.3.8)$$

where

$$I = \frac{1}{(2k+3)!} \cdot \overline{V}_{n,2k+1,\alpha}\left((t-x)^{2k+3} \cdot R(t, x), x\right).$$

From (3.3.4), we get

$$|\overline{V}_{n,2k+1,\alpha}(\psi_x^{2k+3}, x)| = O\left(n^{-[\frac{2k+4}{2}]}\right) = O\left(n^{-(k+2)}\right). \qquad (3.3.9)$$

Let $\varepsilon > 0$ be given. Since $R(t, x) \to 0$ as $t \to x$, there exists a $\delta > 0$ such that when $|t - x| < \delta$ we have $|R(t, x)| < \varepsilon$ and when $|t - x| \geq \delta$ we have

$$|R(t, x)| \leq C < C \cdot \frac{(t-x)^2}{\delta^2}.$$

Thus for all $t, x \in [0, \infty)$

$$|R(t, x)| \leq \varepsilon + C \cdot \frac{(t-x)^2}{\delta^2}$$

and

$$|I| \leq C\varepsilon \cdot n^{-(k+2)} + \frac{C}{\delta^2} \cdot \left|B_{n,2k+1,\alpha}((t-x)^{2k+5}, x)\right|$$

$$\leq C\varepsilon \cdot n^{-(k+2)} + \frac{C}{\delta^2} \cdot n^{-(k+3)},$$

implying

$$\lim_{n \to \infty} n^{k+2} \cdot |I| = 0.$$

This completes the proof, by combining the estimates (3.3.7), (3.3.8) and (3.3.9). ∎

3.4 Modulus of Continuity and Lupaş-Beta Operators

Let

$$C_{x^2}[0, \infty) = C[0, \infty) \cap B_{x^2}[0, \infty),$$

where $B_{x^2}[0, \infty)$ is the set of all functions f defined on $\mathbb{R}^+$ satisfying the condition

$$|f(x)| \le M_f (1 + x^2)$$

with some constant M_f, depending only on f, but independent of x. By $C_{x^2}^k[0, \infty)$, we denote the subspace of all continuous functions $f \in B_{x^2}[0, \infty)$ for which

$$\lim_{x \to \infty} \frac{f(x)}{1 + x^2}$$

is finite. The weighted modulus of continuity $\Omega(f, \delta)$ defined on $\mathbb{R}^+ = [0, \infty)$ (see [8]) is given by

$$\Omega(f, \delta) = \sup_{|h| < \delta, \, x \in \mathbb{R}^+} \frac{|f(x + h) - f(x)|}{(1 + h^2)(1 + x^2)} \quad \text{for each } f \in C_{x^2}[0, \infty).$$

We now present the following quantitative Voronovskaja type asymptotic formula:

Theorem 3.7 ([71]) *Let* $f'' \in C_{x^2}^k[0, \infty)$, *and* $x > 0$. *Then, we have*

$$\left| \overline{Y}_n(f, x) - f(x) - \frac{x(x+3)}{2(n-1)} f''(x) \right| \le 8(1 + x^2) \, \mathcal{O}(n^{-1}) \, \Omega\left(f'', \frac{1}{\sqrt{n}}\right).$$

By $C_B[0, \infty)$, we denote the class of all real valued continuous and bounded functions f on $[0, \infty)$. The second order Ditzian-Totik modulus of smoothness is defined by:

$$\omega_\varphi^2(f, \delta) = \sup_{0 \le h \le \delta} \sup_{x \pm h\varphi(x) \in [0, \infty)} |f(x + h\varphi(x)) - 2f(x) + f(x - h\varphi(x))|,$$

$\varphi(x) = \sqrt{x(x+3)}$, $x \ge 0$. The corresponding K-functional is:

$$K_{2,\varphi}(f, \delta^2) = \inf_{h \in W_\infty^2(\varphi)} \{\|f - h\| + \delta^2 \|\varphi^2 h''\|\},$$

where

$$W_\infty^2(\varphi) = \{h \in C_B[0, \infty) : h' \in AC_{loc}[0, \infty) : \varphi^2 h'' \in C_B[0, \infty)\}.$$

By Theorem 2.1.1 of [32], it follows that

$$C^{-1}\omega_\varphi^2(f, \delta) \leq K_{2,\varphi}(f, \delta^2) \leq C\omega_\varphi^2(f, \delta),$$

for some absolute constant $C > 0$.

Theorem 3.8 ([71]) *If $f \in C_B[0, \infty)$ and $n \in \mathbb{N}$, then we have the following inequality:*

$$\|\overline{Y}_n(f, x) - f(x)\| \leq 4\omega_\varphi^2\left(f, \frac{1}{\sqrt{n}}\right).$$

Păltănea in [115] considered the weighted modulus of continuity $\omega_\varphi(f; h)$:

$$\omega_\varphi(f; h) = \sup\left\{|f(x) - f(y)| : x \geq 0, y \geq 0, |x - y| \leq h\varphi\left(\frac{x+y}{2}\right)\right\}, h \geq 0$$

where

$$\varphi(x) = \frac{\sqrt{x}}{1 + x^m}, \ x \in [0, \infty), m \in \mathbb{N}, \ m \geq 2.$$

We consider here those functions, for which we have the property

$$\lim_{h \to 0} \omega_\varphi(f; h) = 0.$$

By E, we denote the subspace of $C[0, \infty)$ which contains the polynomials.

Let us denote by $W_\varphi[0, \infty)$ the subspace of all real functions defined on $[0, \infty)$, for which the two conditions mentioned above hold true.

Also, we can write

$$\overline{Y}_n\left(\left[1 + \left(x + \frac{|t - x|}{2}\right)^m\right]^2; x\right)$$

$$= 1 + 2\sum_{k=0}^{m}\binom{m}{k}x^k\overline{Y}_n(|t - x|^{m-k}, x) \cdot \frac{1}{2^{m-k}}$$

$$+ \sum_{k=0}^{2m}\binom{2m}{k}x^k\overline{Y}_n(|t - x|^{2m-k}, x)\frac{1}{2^{2m-k}}$$

$$= A_{n,m,x}. \tag{3.4.1}$$

It is easy to verify that for fixed x and m, the term $A_{n,m,x}$ defined in (3.4.1) is bounded when $n \to \infty$.

Gupta–Rassias–Pandey [71] applied Theorems 2.2 and 2.3 of [135] and obtained the following results:

Theorem 3.9 ([71]) *If $f \in C^2[0, \infty) \cap E$ and $f'' \in W_\varphi[0, \infty)$, then we have for $x \in (0, \infty)$ that*

$$\left| \overline{Y}_n(f; x) - f(x) - \frac{x(x+3)}{2(n-1)} f''(x) \right| \le \frac{1}{2} \left[\frac{x(x+3)}{n-1} + \sqrt{2A_{n,m,x}} \right]$$

$$.\omega_\varphi \left(f''; \sqrt{\frac{\mu_{n,6}(x)}{x}} \right),$$

where $A_{n,m,x}$ is given by (3.4.1) and

$$\mu_{n,6}(x) = \overline{Y}_n((e_1 - xe_0)^6, x)$$

$$= \frac{1}{(n-1)(n-2)(n-3)(n-4)(n-5)} \Big[(15n^2 + 430n + 600)x^6$$

$$+ (135n^2 + 3870n + 5400)x^5 + (405n^2 + 11740n + 16800)x^4$$

$$+ (405n^2 + 13950n + 27000)x^3 + (6010n + 22320)x^2 + 7560x \Big].$$

Theorem 3.10 ([71]) *If $f \in C^2[0, \infty) \cap E$ and $f'' \in W_\varphi[0, \infty)$, then we have for $x \in (0, \infty)$ that*

$$\left| (n-1) \left[\overline{Y}_n(f; x) - f(x) - \frac{x(x+3)}{n-1} f''(x) \right] \right|$$

$$\le \frac{1}{2} \left[x(x+3) + \sqrt{2}\sqrt{x}(x+3)C_{n,2,m}(x) \right] \omega_\varphi \left(f; \sqrt{\frac{\mu_{n,4}(x)}{\mu_{n,2}(x)}} \right),$$

where

$$\mu_{n,2}(x) = \overline{Y}_n((e_1 - xe_0)^2, x) = \frac{x(x+3)}{n-1},$$

$$\mu_{n,4}(x) = \overline{Y}_n((e_1 - xe_0)^4, x)$$

$$= \frac{3(n+6)x^4 + 18(n+6)x^3 + (27n + 168)x^2 + 90x}{(n-1)(n-2)(n-3)}.$$

and

$$C_{n,2,m}(x) = 1 + \frac{1}{\overline{Y}_n(|t-x|^3,x)} \sum_{k=0}^{m} \binom{m}{k} x^{m-k} \frac{\overline{Y}_n(|t-x|^{k+3},x)}{2^k}.$$

For the operators $\overline{Y}_n$, we assume that

$$\frac{\overline{Y}_n(|t-x|^k,x)}{\overline{Y}_n(|t-x|^3,x)}, \; 4 \leq k \leq m$$

is a bounded ratio for fixed x and m, when $n \to \infty$.

Remark 3.3 By the same arguments as in [135] the terms $C_{n,2,m}(x)$, $A_{n,m,x}$ are bounded for fixed x and m, when $n \to \infty$.

3.5   Approximation by Modified Kantorovich Operators

For the Bernstein operators defined in (1.1.1), we have

$$B_n(f,x) = \sum_{k=0}^{n} p_{n,k}(x) \, f\left(\frac{k}{n}\right), \quad x \in [0,1],$$

where

$$p_{n,k}(x) = \binom{n}{k} x^k (1-x)^{n-k}, \quad \text{and} \quad p_{n,k}(x) = 0, \text{ if } k < 0 \text{ or } k > n.$$

It is known that

$$p_{n,k}(x) = (1-x)\, p_{n-1,k}(x) + x\, p_{n-1,k-1}(x), \quad 0 < k < n. \tag{3.5.1}$$

The order of approximation of Bernstein operators was intensively studied in the past few decades. Popoviciu in [119, 120] gave a solution of this problem in terms of first order modulus of continuity. An asymptotic error term of the Bernstein operators was first established by Voronovskaja [139]. Very recently Arab et al. in [16] have introduced modified Bernstein operators to improve the degree of approximation as follows:

$$B_n^{M,1}(f,x) = \sum_{k=0}^{n} p_{n,k}^{M,1}(x) \, f\left(\frac{k}{n}\right), \quad x \in [0,1], \tag{3.5.2}$$

$$p_{n,k}^{M,1}(x) = a(x,n)\, p_{n-1,k}(x) + a(1-x,n)\, p_{n-1,k-1}(x)$$

and

$$a(x, n) = a_1(n)\, x + a_0(n),\ n = 0, 1, \ldots,$$

where $a_0(n)$ and $a_1(n)$ are two unknown sequences which are determined appropriately. For $a_1(n) = -1$, $a_0(n) = 1$, clearly, $p_{n,k}^{M,1}(x)$ reduces to (3.5.1). Very recently Gupta et al. in [82] extended the results of Arab et al. [16] and proposed the following classical Kantorovich operators

$$K_n^{M,1}(f, x) = (n + 1) \sum_{k=0}^{n} p_{n,k}^{M,1}(x) \int_{\frac{k}{n+1}}^{\frac{k+1}{n+1}} f(t)dt, \qquad (3.5.3)$$

where $p_{n,k}^{M,1}(x)$ are defined as in (3.5.2).

It was calculated in [82] that

$$K_n^{M,1}(e_0; x) = 2a_0(n) + a_1(n);$$

In order to study the uniform convergence, it has been considered in [82] that sequences $a_i(n)$, $i = 0, 1$ verify the condition

$$2a_0(n) + a_1(n) = 1. \qquad (3.5.4)$$

They considered the following two cases for unknown sequences $a_0(n)$ and $a_1(n)$:
Case 1. Let
$$a_0(n) \geq 0,\ a_0(n) + a_1(n) \geq 0. \qquad (3.5.5)$$

Using condition (3.5.4) they obtained $0 \leq a_0(n) \leq 1$ and $-1 \leq a_1(n) \leq 1$, namely the sequences are bounded. In this case the operator (3.5.3) is positive.
Case 2. Let
$$a_0(n) < 0 \text{ or } a_1(n) + a_0(n) < 0. \qquad (3.5.6)$$

If $a_0(n) < 0$, then $a_1(n) + a_0(n) > 1$ and if $a_1(n) + a_0(n) < 0$, then $a_0(n) > 1$. In this case the operator (3.5.3) is not positive.

Theorem 3.11 ([82]) *Let $a_1(n)$, $a_0(n)$ be two sequences which verify the conditions (3.5.4) and (3.5.5). If $f \in C[0, 1]$. Then*

$$\lim_{n \to \infty} K_n^{M,1}(f; x) = f(x),$$

uniformly on $[0, 1]$.

The above result can be extended for the Case 2. In order to prove this result one can recall the extended form of Korovkin's Theorem.

Theorem 3.12 ([16, Theorem 10]) *Let* $0 < h \in C[a, b]$ *be a function and suppose that* $(L_n)_{n \geq 1}$ *be a sequence of positive linear operators such that* $\lim_{n \to \infty} L_n(e_i) = he_i, i = 0, 1, 2,$ *uniformly on* [a,b]. *Then for a given function* $f \in C[a, b]$ *we have* $\lim_{n \to \infty} L_n(f) = hf$ *uniformly on* [a, b].

Theorem 3.13 ([82]) *Let* $f \in C[0, 1]$. *Then for all bounded sequences* $a_1(n)$ *and* $a_0(n)$ *that satisfy the conditions (3.5.4) and (3.5.6), we have*

$$\lim_{n \to \infty} K_n^{M,1}(f; x) = f(x),$$

uniformly on [0, 1].

Theorem 3.14 ([82]) *Let* $a_i(n)$ *be a convergent sequence that satisfies the conditions (3.5.4), (3.5.5) and* $l_i = \lim_{n \to \infty} a_i(n), i = 0, 1$. *If* $f'' \in C[0, 1]$, *then*

$$\lim_{n \to \infty} n \left(K_n^{M,1}(f; x) - f(x) \right) = \frac{1}{2}(1 - 2x)(3l_1 + 4l_0)f'(x) + \frac{1}{2}x(1 - x)(2l_0 + l_1)f''(x),$$

uniformly on [0, 1].

Theorem 3.15 ([82]) *Let* $a_i(n), i = 0, 1$ *be bounded convergent sequences which satisfy (3.5.4), (3.5.6) and* $l_i = \lim_{n \to \infty} a_i(n), i = 0, 1$. *If* $f \in C[0, 1]$ *and* f'' *exists at a certain point* $x \in [0, 1]$, *then we have*

$$\lim_{n \to \infty} n[K_n^{M,1}(f; x) - f(x)] = \frac{1}{2}(1 - 2x)(3l_1 + 4l_0)f'(x)$$

$$+ \frac{1}{2}x(1 - x)(2l_0 + l_1)f''(x). \tag{3.5.7}$$

Moreover the relation (3.5.7) holds uniformly on [0, 1] *if* $f'' \in C[0, 1]$.

Remark 3.4 It was pointed out in [82] that the proofs of Voronovskaja-type estimates [16, Theorems 8 and 13] are not correct because the relevant upper bounds in the proof of the above theorem for modified Bernstein operators when using only second central moment (instead of fourth central moment) are not enough.

Theorem 3.16 ([82]) *If* $f(x)$ *is bounded for* $x \in [0, 1]$, $a_0(n)$, $a_1(n)$ *satisfy (3.1.4) and* $a_1(n)$ *is a bounded sequence, then*

$$\|K_n^{M,1}f - f\| \leq (3|a_1(n)| + 1)\left(1 + \frac{1}{\sqrt{6}}\right) \omega\left(f; \frac{1}{\sqrt{n}}\right),$$

where $\| \cdot \|$ *is the uniform norm on the interval* [0, 1] *and* $\omega(f, \delta)$ *is the first order modulus of continuity.*

Theorem 3.17 ([82]) *For* $g \in C^2[0, 1]$, $x \in [0, 1]$ *fixed and* $K_n^{M,1}$ *defined as above, we have*

$$\left| K_n^{M,1}(g; x) - g(x) - \frac{1}{2} K_n^{M,1}\left((t - x)^2; x\right) g''(x) \right| \leq C \frac{1}{n} \omega \left(g'', \frac{1}{\sqrt{n}} \right),$$

where $C > 0$ *is a constant independent of* n, x.

Corollary 3.4 ([82]) *For* $g \in C^2[0, 1]$, $x \in [0, 1]$ *fixed, we have*

$$\lim_{n \to \infty} n \left[K_n^{M,1}(g; x) - g(x) \right] = \frac{x(1 - x)}{2} g''(x).$$

Corollary 3.5 ([82]) *For* $f \in C^2[0, 1]$ *the following holds true*

$$\| K_n^{M,1} g - g \|_{C[0,1]} \leq \frac{C}{n} \| g'' \|.$$

Theorem 3.18 ([82]) *For* $f \in C[0, 1]$, $a_1(n) = -2$, $a_0(n) = \frac{3}{2}$ *the following holds true*

$$\| K_n^{M,1} f - f \|_{C[0,1]} \leq C \omega_2 \left(f; \frac{1}{\sqrt{n}} \right). \tag{3.5.8}$$

Remark 3.5 Clearly Theorem 3.18 is better than Theorem 3.16 since now we have an estimate in terms of ω_2 instead of ω_1. Also, we observe that for e_0, e_1, e_2 in place of $g(x)$ in Theorem 3.17 in both sides of the inequality we get 0.

Remark 3.6 [82] According to Voronovskaja's Theorem, the optimal rate of approximation for the class $C[0, 1]$ is exactly $O\left(\frac{1}{n}\right)$ independently of how smooth is the approximated function f (see [29, Theorem 5.1]). The same optimal rate of approximation (saturation order) $\frac{1}{n}$ is valid for the Kantorovich operator (see [29, Theorem 6.3, p. 317]).

In the modified Kantorovich operator considered in [82], when $a_1(n) = -2$, $a_0(n) = \frac{3}{2}$ we derive from Corollary 3.4 an order of approximation better than $\frac{1}{n}$ for $f \in C^2[0, 1]$. Furthermore, if we suppose $g \in C^3[0, 1]$, using

$$\omega \left(g'', \frac{1}{\sqrt{n}} \right) \leq \frac{1}{\sqrt{n}} \| g''' \|_{C[0,1]},$$

from Theorem 3.17 we obtain the proof of the inequality (3.5.9).

Corollary 3.6 ([82]) *If* $g \in C^3[0, 1]$, *then*

$$\lim_{n \to \infty} n^{\frac{3}{2}} \left| K_n^{M,1}(g; x) - g(x) \right| \leq C \| g''' \|_{C[0,1]}, \tag{3.5.9}$$

where $C > 0$ is a constant independent of n and x.

Also for the second modification of the Kantorovich operator $K_n^{M,2}$, the order of approximation $O\left(\dfrac{1}{n^2}\right)$ was obtained in [82]. But, this case lacks the positivity of the operators.

Another modification of the Kantorovich operators considered in [82] is the following:

$$K_n^{M,2}(f;x) = (n+1) \sum_{k=0}^{n} p_{n,k}^{M,2}(x) \int_{\frac{k}{n+1}}^{\frac{k+1}{n+1}} f(t)dt, \qquad (3.5.10)$$

where

$$p_{n,k}^{M,2}(x) = b(x,n)\,p_{n-2,k}(x) + d(x,n)\,p_{n-2,k-1}(x) + b(1-x,n)\,p_{n-2,k-2}(x) \qquad (3.5.11)$$

and

$$b(x,n) = b_2(n)x^2 + b_1(n)x + b_0(n), \quad d(x,n) = d_0(n)x(1-x),$$

where $b_i(n)$, $i = 0, 1, 2$ and $d_0(n)$ are two unknown sequences which are determined appropriately. For $b_2(n) = b_0(n) = 1$, $b_1(n) = -2$, $d_0(n) = 2$, (3.5.11) clearly reduces to the Bernstein basis defined in (1.1.1). In [82] it was calculated that

$$K_n^{M,2}(e_0;x) = (2b_2(n) - d_0(n))x^2 - (2b_2(n) - d_0(n))x + b_2(n) + 2b_0(n) + b_1(n);$$

In order to study the uniform convergence we set $K_n^{M,2}(e_0;x) = 1$, and this yields:

$$2b_2(n) - d_0(n) = 0, \quad b_2(n) + 2b_0(n) + b_1(n) = 1.$$

Using the above relations we obtain

$$K_n^{M,2}(e_1;x) = x + \frac{(2x-1)(4b_0(n)-5)}{2(n+1)};$$

$$K_n^{M,2}(e_2;x) = x^2 + \frac{1}{3(n+1)^2}\left\{(24b_0(n)x^2 - 12b_0(n)x - 33x^2 + 18x)n\right.$$

$$\left. -60b_0(n)x^2 - 6b_1(n)x^2 + 72b_0(n)x + 6b_1(n)x + 45x^2 - 18b_0(n) - 60x + 19\right\}.$$

In order to have

$$\lim_{n\to\infty} K_n^{M,2}(e_i;x) = x^i, \quad i = 0, 1, 2,$$

we consider the sequences $b_0(n)$ and $b_1(n)$ to verify the conditions

$$\lim_{n\to\infty} \frac{b_0(n)}{n} = 0 \text{ and } \lim_{n\to\infty} \frac{b_1(n)}{n^2} = 0.$$

We propose our analysis for the case

$$bo(n) = \frac{5}{4} \text{ and } b_1(n) = -\frac{n}{2},$$

therefore

$$b_2(n) = \frac{n-3}{2} \text{ and } d_0(n) = n - 3.$$

With the above choices the operator (3.5.10) becomes

$$\tilde{K}_n^{M,2}(f; x) = (n+1) \sum_{k=0}^{n} \tilde{p}_{n,k}^{M,2}(x) \int_{\frac{k}{n+1}}^{\frac{k+1}{n+1}} f(t)dt, \qquad (3.5.12)$$

where

$$\tilde{p}_{n,k}^{M,2}(x) = \left(\frac{n-3}{2}x^2 - \frac{n}{2}x + \frac{5}{4}\right) p_{n-2,k}(x) + (n-3)x(1-x)p_{n-2,k-1}(x)$$
$$+ \left(\frac{n-3}{2}x^2 + \frac{6-n}{2}x - \frac{1}{4}\right) p_{n-2,k-2}(x).$$

Note that other choices for sequences $b_0(n)$ and $b_1(n)$ lead to some operators with order of approximation either one or two. It was pointed out in [82] that the operators $\tilde{K}_n^{M,2}$ preserve constant and linear functions.

Theorem 3.19 ([82]) *If $f \in C^6[0, 1]$ and $x \in [0, 1]$, then for sufficiently large n, we have*

$$|\tilde{K}_n^{M,2}(f; x) - f(x)| = \mathcal{O}\left(\frac{1}{n^2}\right).$$

The following Kantorovich operators considered in [82], constitute another approach:

$$K_n^{M,3}(f; x) = (n+1) \sum_{k=0}^{n} p_{n,k}^{M,3}(x) \int_{\frac{k}{n+1}}^{\frac{k+1}{n+1}} f(t)dt, \qquad (3.5.13)$$

where

$$p_{n,k}^{M,3}(x) = \tilde{b}(x, n)p_{n-4,k}(x) + \tilde{d}(x, n)p_{n-4,k-1}(x) + \tilde{e}(x, n)p_{n-4,k-2}(x)$$
$$+ \tilde{d}(1-x, n)p_{n-4,k-3}(x) + \tilde{b}(1-x, n)p_{n-4,k-4}(x)$$

and

$$\tilde{b}(x, n) = \tilde{b}_4(n)x^4 + \tilde{b}_3(n)x^3 + \tilde{b}_2(n)x^2 + \tilde{b}_1(n)x + \tilde{b}_0(n),$$
$$\tilde{d}(x, n) = \tilde{d}_4(n)x^4 + \tilde{d}_3(n)x^3 + \tilde{d}_2(n)x^2 + \tilde{d}_1(n)x + \tilde{d}_0(n),$$
$$\tilde{e}(x, n) = \tilde{e}_0(n)(x(1-x))^2.$$

We note that $b_i(n)$, $d_i(n)$, $i = 0, 1, \ldots, 4$ and $e_0(n)$ are some unknown sequences which are determined appropriately. Let $\tilde{K}_n^{M,3}$ be the operator (3.5.13) with the following sequences:

$$\tilde{b}_0(n) = \frac{137}{72}, \ \tilde{b}_1(n) = -\frac{69}{8} - \frac{17}{24}n, \ \tilde{b}_2(n) = \frac{n^2}{8},$$

$$\tilde{b}_3(n) = \frac{115}{6} + \frac{5}{2}n - \frac{1}{4}n^2, \ \tilde{b}_4(n) = -\frac{101}{8} - \frac{43}{24}n + \frac{1}{8}n^2,$$

$$\tilde{d}_0(n) = -\frac{43}{36}, \ \tilde{d}_1(n) = \frac{45}{4} + \frac{5}{4}n,$$

$$\tilde{d}_2(n) = \frac{115}{4} - \frac{1}{2}n^2 + \frac{15}{4}n, \ \tilde{d}_3(n) = -\frac{533}{6} + n^2 - \frac{73}{6}n,$$

$$\tilde{d}_4(n) = \frac{101}{2} + \frac{43}{6}n - \frac{1}{2}n^2, \ \tilde{e}_0(n) = -\frac{303}{4} - \frac{43}{4}n + \frac{3}{4}n^2.$$

Theorem 3.20 ([82]) *If $f \in C^{10}[0, 1]$ and $x \in [0, 1]$, then for sufficiently large n, we have*

$$|\tilde{K}_n^{M,3}(f; x) - f(x)| = \mathcal{O}\left(\frac{1}{n^3}\right).$$

3.6 Gamma Transform and Convergence

Miheşan in [106] defined the gamma transform of a function f as

$$(\Gamma_{\alpha,\beta}^{(a)} f)(x) = \frac{\beta^\alpha}{\Gamma(\alpha)} \int_0^\infty t^{\alpha-1} e^{-\beta t} f(t^a) dt, \tag{3.6.1}$$

where $a \in \mathbb{R}$, $\alpha, \beta > 0$ and $f \in L_{1,loc}(0, \infty)$.

Case I When $a = 1$, clearly $\Gamma_{\alpha,\beta}^{(a)} e_1 = \alpha/\beta$. In order to preserve the test function e_1, Miheşan [106] considered $\alpha/\beta = x$ in (3.6.1) and obtained the following form of linear positive operators

$$(\Gamma_\alpha f)(x) = \left(\frac{\alpha}{x}\right)^\alpha \frac{1}{\Gamma(\alpha)} \int_0^\infty t^{\alpha-1} e^{-\frac{\alpha t}{x}} f(t) dt. \tag{3.6.2}$$

For the special case $\alpha = n$, (3.6.2) reduces to the Post-Widder operators defined as

$$(P_n f)(x) = \left(\frac{n}{x}\right)^n \frac{1}{\Gamma(n)} \int_0^\infty t^{n-1} e^{-\frac{nt}{x}} f(t) dt. \tag{3.6.3}$$

Additionally, for another special case, when $\alpha = nx$ then (3.6.2) reduces to the Rathore operators defined as

$$(R_n f)(x) = \frac{n^{nx}}{\Gamma(nx)} \int_0^\infty t^{nx-1} e^{-nt} f(t) dt. \tag{3.6.4}$$

Furthermore, it was shown in [106], that if we consider $(R_n)(S_n f)$, where S_n stands for the Szász-Mirakyan operators defined in (1.3.1), we get the Lupaş operators

$$(L_n f)(x) := (R_n)(S_n f) = 2^{-nx} \sum_{k=0}^\infty \frac{(nx)^k}{2^k . k!} f\left(\frac{k}{n}\right). \tag{3.6.5}$$

Case II When $a = -1$ then clearly $\Gamma_{\alpha,\beta}^{(a)} e_1 = \beta/(\alpha - 1)$. In order to preserve the test function e_1, Miheşan [106] considered $\beta/(\alpha - 1) = x$ in (3.6.1) and obtained the following form of linear positive operators

$$(\tilde{\Gamma}_\alpha f)(x) = \frac{(\alpha - 1)^\alpha x^\alpha}{\Gamma(\alpha)} \int_0^\infty t^{\alpha-1} e^{-(\alpha-1)xt} f\left(\frac{1}{t}\right) dt. \tag{3.6.6}$$

For the special case $\alpha = n + 1$, (3.6.6) reduces to the Lupaş-Müller operators defined as

$$(G_n f)(x) = \frac{(nx)^{n+1}}{\Gamma(n + 1)} \int_0^\infty t^n e^{-ntx} f\left(\frac{1}{t}\right) dt. \tag{3.6.7}$$

For $x \geq 0$ and $\alpha \in \mathbb{R}$, and taking the Γ_α transform of the Szász-Mirakyan operators S_n, Miheşan in [106] proposed the following operators

$$M_{n,\alpha}(f, x) := (\Gamma_\alpha)(S_n f) = \sum_{k=0}^\infty \frac{(\alpha)_k}{k!} \frac{\left(\frac{nx}{\alpha}\right)^k}{\left(1 + \frac{nx}{\alpha}\right)^{\alpha+k}} f\left(\frac{k}{n}\right), \tag{3.6.8}$$

where $(\alpha)_k = \alpha(\alpha + 1)(\alpha + 2) \cdots (\alpha + k - 1)$, $(\alpha)_0 = 1$.

In terms of Miheşan's basis function, the Durrmeyer operators can be defined as follows

$$D_{n,\alpha}(f, x) = \frac{n(\alpha - 1)}{\alpha} \sum_{k=0}^\infty m_{n,k}^\alpha(x) \int_0^\infty m_{n,k}^\alpha(t) f(t) dt, \tag{3.6.9}$$

where

$$m_{n,k}^\alpha(x) = \frac{(\alpha)_k}{k!} \frac{\left(\frac{nx}{\alpha}\right)^k}{\left(1 + \frac{nx}{\alpha}\right)^{\alpha+k}}.$$

We note here that n/α is a finite quantity. We have the following special cases:

1. If $\alpha \to \infty$ we get the Szász-Durrmeyer operators
2. If $\alpha = n$, we get the Baskakov-Durrmeyer operators
3. If $\alpha = -n$, we get the Bernstein-Durrmeyer polynomials

The r-th order moments are given by

$$
\begin{aligned}
D_{n,\alpha}(e_r, x) &= \frac{n(\alpha - 1)}{\alpha} \sum_{k=0}^{\infty} m_{n,k}^{\alpha}(x) \int_0^{\infty} m_{n,k}^{\alpha}(t) t^r \, dt \\
&= \frac{n(\alpha - 1)}{\alpha} \sum_{k=0}^{\infty} m_{n,k}^{\alpha}(x) \int_0^{\infty} \frac{(\alpha)_k}{k!} \cdot \frac{\left(\frac{nt}{\alpha}\right)^k}{\left(1 + \frac{nt}{\alpha}\right)^{\alpha+k}} t^r \, dt \\
&= \frac{n(\alpha - 1)}{\alpha} \sum_{k=0}^{\infty} m_{n,k}^{\alpha}(x) \frac{(\alpha)_k}{k!} \cdot B(k + r + 1, \alpha - r - 1) \left(\frac{\alpha}{n}\right)^{r+1} \\
&= \frac{n(\alpha - 1)}{\alpha} \cdot \left(\frac{\alpha}{n}\right)^{r+1} \sum_{k=0}^{\infty} m_{n,k}^{\alpha}(x) \frac{\Gamma(k + \alpha)}{\Gamma(\alpha).k!} \cdot \frac{\Gamma(k + r + 1)\Gamma(\alpha - r - 1)}{\Gamma(\alpha + k)} \\
&= \frac{n(\alpha - 1)}{\alpha} \cdot \Gamma(\alpha - r - 1) \cdot \left(\frac{\alpha}{n}\right)^{r+1} \sum_{k=0}^{\infty} m_{n,k}^{\alpha}(x) \frac{\Gamma(k + r + 1)}{\Gamma(\alpha).k!} \\
&= \frac{n(\alpha - 1)}{\alpha} \cdot \Gamma(\alpha - r - 1) \cdot \left(\frac{\alpha}{n}\right)^{r+1} \sum_{k=0}^{\infty} \frac{(\alpha)_k}{k!} \frac{\left(\frac{nx}{\alpha}\right)^k}{\left(1 + \frac{nx}{\alpha}\right)^{\alpha+k}} \frac{\Gamma(k + r + 1)}{\Gamma(\alpha).k!} \\
&= \frac{\Gamma(\alpha - r - 1)\Gamma(r + 1)}{\Gamma(\alpha - 1)} \left(\frac{\alpha}{n}\right)^r \sum_{k=0}^{\infty} \frac{(\alpha)_k (r + 1)_k}{k!(1)_k} \frac{\left(\frac{nx}{\alpha}\right)^k}{\left(1 + \frac{nx}{\alpha}\right)^{\alpha+k}} \Gamma(r + 1) \\
&= \frac{\Gamma(\alpha - r - 1)\Gamma(r + 1)}{\Gamma(\alpha - 1)} \left(\frac{\alpha}{n}\right)^r \left(1 + \frac{nx}{\alpha}\right)^{-\alpha} {}_2F_1\left(\alpha, r + 1; 1; \frac{nx}{nx + \alpha}\right)
\end{aligned}
$$

Applying Kummer's transformation

$$
{}_2F_1(a, b; c; z) = (1 - z)^{-a} \, {}_2F_1\left(a, c - b; c; \frac{z}{z - 1}\right).
$$

we get

$$
D_{n,\alpha}(e_r, x) = \frac{\Gamma(\alpha - r - 1)\Gamma(r + 1)}{\Gamma(\alpha - 1)} \left(\frac{\alpha}{n}\right)^r {}_2F_1\left(\alpha, -r; 1; \frac{-nx}{\alpha}\right).
$$

Remark 3.7 By a simple computation, we have

$$
D_{n,\alpha}(e_0, x) = 1
$$

$$
D_{n,\alpha}(e_1, x) = \frac{\alpha(1 + nx)}{n(\alpha - 2)}
$$

$$
D_{n,\alpha}(e_2, x) = \frac{\alpha[n^2(1 + \alpha)x^2 + 4n\alpha x + 2\alpha]}{n^2(\alpha - 2)(\alpha - 3)}.
$$

Remark 3.8 By a simple computation, we have

$$D_{n,\alpha}((e_1 - xe_0), x) = \frac{\alpha + 2nx}{n(\alpha - 2)}$$

$$D_{n,\alpha}((e_1 - xe_0)^2, x) = \frac{2n^2x^2(\alpha + 3) + 2n\alpha x(\alpha + 3) + 2\alpha^2}{n^2(\alpha - 2)(\alpha - 3)}.$$

Theorem 3.21 *Let* $f \in C_B[0, \infty)$, *then for* $r \in \mathbb{N}$, *we have*

$$\left| D_{n,\alpha}(f, x) - f(x) \right| \leq C\omega_2 \left(f, \sqrt{\delta_{n,\alpha}} \right) + \omega \left(f, \left| \frac{\alpha + 2nx}{n(\alpha - 2)} \right| \right)$$

where C is a positive constant and $\delta_{n,\alpha}$ *is given as:*

$$\delta_{n,\alpha} = \frac{x^2(2n^2\alpha^2 + 6n^2\alpha - 2n^2) + x(2n\alpha^3 + 6n\alpha^2 - 24n\alpha) + 3\alpha^3 - 7\alpha^2}{n^2(\alpha - 2)^2(\alpha - 3)}$$

Here have convergence if α *and n both tend to* ∞.

Proof We introduce the auxiliary operators

$$\tilde{D}_{n,\alpha}(f, x) : C_B[0, \infty) \to C_B[0, \infty)$$

as follows

$$\tilde{D}_{n,\alpha}(f, x) = D_{n,\alpha}(f, x) - f\left(\frac{\alpha(1 + nx)}{n(\alpha - 2)} \right) + f(x). \qquad (3.6.10)$$

These operators preserve linear functions. By Remark 3.7, we derive that

$$\tilde{D}_{n,\alpha}(e_1, x) = D_{n,\alpha}(e_1, x) - \frac{\alpha(1 + nx)}{n(\alpha - 2)} + x = x.$$

Let $h \in C_B^2[0, \infty)$ and $x, t \in [0, \infty)$. By Taylor's formula, we obtain

$$h(t) = h(x) + (t - x)g'(x) + \int_x^t (t - u)h''(u)du.$$

Then using Remark 3.8, we have

$$\left| \tilde{D}_{n,\alpha}(h, x) - h(x) \right| = \left| \tilde{D}_{n,\alpha} \left(\int_x^t (t - u)h''(u)du, x \right) \right|$$

$$= \left| D_{n,\alpha} \left(\int_x^t (t - u)h''(u)du, x \right) \right|$$

$$
-\int_x^{\frac{\alpha(1+nx)}{n(\alpha-2)}}\left(\frac{\alpha(1+nx)}{n(\alpha-2)}-u\right)h''(u)du\bigg|
$$

$$
\leq\left[D_{n,\alpha}((e_1-xe_0)^2,x)+\left(\frac{\alpha(1+nx)}{n(\alpha-2)}-x\right)^2\right]||h''||
$$

$$
:=\delta_{n,\alpha}||h''||. \tag{3.6.11}
$$

Subsequently, by (3.6.10) and Remark 3.7, we have

$$
\left|\widetilde{D}_{n,\alpha}(f,x)\right|\leq D_{n,\alpha}(1,x)||f||+2||f||\leq 3||f||. \tag{3.6.12}
$$

Using (3.6.10), (3.6.11) and (3.6.12), we get

$$
\left|D_{n,\alpha}(f,x)-f(x)\right|\leq\left|\widetilde{D}_{n,\alpha}(f-h,x)-(f-h)(x)\right|+\left|\widetilde{D}_{n,\alpha}(h,x)-h(x)\right|
$$

$$
+\left|f\left(\frac{\alpha(1+nx)}{n(\alpha-2)}\right)-f(x)\right|
$$

$$
\leq 4||f-g||+\delta_{n,\alpha}||g''||+\left|f\left(\frac{\alpha(1+nx)}{n(\alpha-2)}\right)-f(x)\right|
$$

$$
\leq C\left\{||f-h||+\delta_{n,\alpha}||h''||\right\}+\omega\left(f,\left|\frac{(\alpha+2nx)}{n(\alpha-2)}\right|\right).
$$

Finally, if we take the infimum over all $h\in C_B^2[0,\infty)$, (see in Sect. 3.2) and apply the well-known inequality due to DeVore-Lorentz [29]

$$
K_2(f,\delta)\leq C\omega_2(f,\sqrt{\delta}),\ \delta>0,
$$

we obtain the desired result.   ∎

Remark 3.9 We note that some other results analogous to [53] can be obtained for $D_{n,\alpha}(f,x)$.

3.7   Difference of Mastroianni and Srivastava-Gupta Operators

Recently Gupta et al. in [68, Chap. 7] studied some problems on the difference of operators. This section deals with a quantitative estimate for the difference of Srivastava-Gupta operators $G_{n,c}$ defined in (2.22.1) and Mastroianni operators [113] defined as

$$
M_{n,c}(f;x)=\sum_{k=0}^{\infty}p_{n,k}(x,c)F_k(f)
$$

where $p_{n,k}(x, c)$ is given in (2.22.1) and $F_k(f) = f(k/n)$.

Alternatively, we may write the Srivastava-Gupta operators (2.22.1) as

$$G_{n,c}(f, x) = \sum_{k=0}^{\infty} p_{n,k}(x, c)G_k(f)$$

where

$$G_k(f) = n \int_0^{\infty} p_{n+c,k-1}(t, c)f(t)dt, \, 1 \le k < \infty, \quad G_0(f) = f(0).$$

For $C_B[0, \infty)$, the class of bounded continuous functions defined on the interval $[0, \infty)$ and with the norm

$$||.|| = \sup_{x \in [0, \infty)} |f(x)| < \infty,$$

the following quantitative estimate was established by Gupta in [55].

Theorem 3.22 *Let* $f^{(s)} \in C_B[0, \infty)$, $s \in \{0, 1, 2\}$ *and* $x \in [0, \infty)$, *then for* $n \in \mathbb{N}$, *we have*

$$|(G_{n,c} - M_{n,c})(f, x)| \le ||f''||\alpha(x) + 2\omega(f, \delta),$$

where

$$\alpha(x) = \frac{cnx^2(n + c) + n^2x}{(n - c)^2(n - 2c)}, \quad \delta = \frac{c\sqrt{x[x(n + c) + 1]}}{\sqrt{n}(n - c)}$$

and $\omega(f, .)$ *is the usual modulus of continuity.*

Remark 3.10 One can study analogous difference estimates for the other new operators proposed in [55, Ex. 2.1–2.4].

3.8 Difference of Two Operators

Let $F : D \to \mathbb{R}$ be a positive linear functional defined on a subspace D of $C[0, \infty)$, containing polynomials up to fourth degree such that $F(e_0) = 1$, $b^F := F(e_1)$, $\mu_r^F = F(e_1 - b^F e_0)^r$, $r \in \mathbb{N}$ with $e_m(t) = t^m$, $m \in \mathbb{N}$.

Let us consider $F_{n,k} : D \to \mathbb{R}$, $G_{n,k} : D \to \mathbb{R}$ and define the operators having same basis $p_{n,k}(x)$ such that

$$U_n(f, x) = \sum_{k=0}^{\infty} p_{n,k}(x)F_{n,k}(f), \quad V_n(f, x) = \sum_{k=0}^{\infty} p_{n,k}(x)G_{n,k}(f).$$

The following quantitative general result for the difference of two operators U_n and V_n was given in [56]:

Theorem 3.23 ([56]) *Let* $f^{(s)} \in C_B[0, \infty), s \in \{0, 1, 2\}$ *and* $x \in [0, \infty)$, *then for* $n \in \mathbb{N}$, *we have*

$$|(U_n - V_n)(f, x)| \leq ||f''||\alpha(x) + \omega(f'', \delta_1)(1 + \alpha(x)) + 2\omega(f, \delta_2),$$

where

$$\alpha(x) = \frac{1}{2} \sum_{k=0}^{\infty} p_{n,k}(x)(\mu_2^{F_{n,k}} + \mu_2^{G_{n,k}})$$

and

$$\delta_1^2 = \frac{1}{2} \sum_{k=0}^{\infty} p_{n,k}(x)(\mu_4^{F_{n,k}} + \mu_4^{G_{n,k}}), \delta_2^2 = \sum_{k=0}^{\infty} p_{n,k}(x)(b^{F_{n,k}} - b^{G_{n,k}})^2,$$

with $b^F := F(e_1)$ *and* $\mu_r^F = F(e_1 - b^F e_0)^r, r \in \mathbb{N}$.

Applying the above theorem, Gupta in [56], considered the differences of Szász operators with the Szász-Baskakov operators, Phillips operators, Szász-Kantorovich operators and Szász-Durrmeyer operators. Additionally, in [57] Gupta obtained further results for the difference of the Lupaş operators and its variants, having same basis $l_{n,k}(x)$:

The Lupaş operators (see [100]), for $f \in C[0, \infty)$ are defined as

$$L_n(f, x) := \sum_{k=0}^{\infty} l_{n,k}(x) F_{n,k}(f), \tag{3.8.1}$$

where $F_{n,k} : D \to \mathbb{R}$ is a positive linear functional defined on a subspace D of $C[0, \infty)$ and

$$l_{n,k}(x) = 2^{-nx} \frac{(nx)_k}{k! 2^k}, \quad F_{n,k}(f) = f\left(\frac{k}{n}\right)$$

We consider Lupaş-Kantorovich operators V_n as

$$K_n(f, x) := \sum_{k=0}^{\infty} l_{n,k}(x) G_{n,k}(f) \tag{3.8.2}$$

where $G_{n,k} : D \to \mathbb{R}$ are defined by

$$G_{n,k}(f) = n \int_{\frac{k}{n}}^{\frac{k+1}{n}} f(t)\, dt.$$

The following difference estimate between Lupaş operators (3.8.1) and Lupaş-Kantorovich operators (3.8.2) was established in [57].

Theorem 3.24 *Let $f^{(s)} \in C_B[0, \infty)$, $s \in \{0, 1, 2\}$ and $x \in [0, \infty)$, then for $n \in \mathbb{N}$, we have*

$$|(K_n - L_n)(f, x)| \leq \frac{1}{24n^2}\|f''\| + \omega\left(f'', \frac{1}{4\sqrt{10}n^2}\right)\left(1 + \frac{1}{24n^2}\right) + 2\omega\left(f, \frac{1}{2n}\right).$$

The Lupaş-Szász operators are defined as

$$S_n(f; x) = \sum_{k=0}^{\infty} l_{n,k}(x) H_{n,k}(f), \tag{3.8.3}$$

where

$$H_{n,k}(f) = n \int_0^{\infty} s_{n,k-1}(t) f(t)\, dt,\ 1 \leq k < \infty,\ H_{n,0}(f) = f(0)$$

and the Szász basis function is defined as

$$s_{n,k}(t) = \frac{e^{-nt}(nt)^k}{k!}.$$

The following difference result between Lupaş operators (3.8.1) and Lupaş-Szász operators (3.8.3) was established in [57]:

Theorem 3.25 *Let $f^{(s)} \in C_B[0, \infty)$, $s \in \{0, 1, 2\}$ and $x \in [0, \infty)$, then for $n \in \mathbb{N}$, we have*

$$|(S_n - L_n)(f, x)| = \|f''\|\frac{x}{2n} + \omega\left(f'', \sqrt{\frac{3x^2}{2n^2} + \frac{6x}{n^3}}\right)\left(1 + \frac{x}{2n}\right).$$

Some general quantitative estimates for the difference of operators, having different basis functions will appear shortly in forthcoming studies by Gupta and collaborators.

References

1. U. Abel, The moment for the Meyer-König and Zeller operators. J. Approx. Theory **82**, 353–361 (1995)
2. U. Abel, B. Della Vecchia, Asymptotic approximation by the operators of K. Balázs and Szabados. Acta Sci. Math. (Szeged) **66**(12), 137–145 (2000)
3. U. Abel, V. Gupta, Rate of convergence of Stancu beta operators for functions of bounded variation, Revue D' Anal. Numerique et de Theorie de L' Approximation **33**(1), 3–9 (2004)
4. U. Abel, V. Gupta, R.N. Mohapatra, Local approximation by beta operators. Nonlinear Anal. **62**(1), 41–52 (2005). (Series A, Theory and Methods)
5. U. Abel, V. Gupta, R.N. Mohapatra, Local approximation by a variant of Bernstein-Durrmeyer operators. Nonlinear Anal. **68**, 3372–3381 (2008)
6. U. Abel, M. Ivan, On a generalization of an approximation operator defined by A. Lupaş, General Math. **15**(1), 21–34 (2007)
7. T. Acar, A. Aral, H. Gonska, On Szász-Mirakyan operators preserving e^{2ax}, $a>0$, Mediterr. J. Math. **14**(1), Article ID 6, 14 (2017)
8. T. Acar, A. Aral, I. Rasa, The new forms of Voronovskaja's theorem in weighted spaces. Positivity **20**, 25–40 (2016)
9. O. Agratini, Approximation properties of a class of linear operators. Math. Method. Appl. Sci. **36**, 2353–2358 (2013)
10. P.N. Agrawal, V. Gupta, Inverse theorem for linear combinations of modified Bernstein polynomials. Pure Appl. Math. Sci. **40**, 29–38 (1995)
11. P.N. Agrawal, V. Gupta, A.S. Kumar, A. Kajla, Generalized Baskakov Szász type operators. Appl. Math. Comput. **236**(1), 311–324 (2014)
12. P.N. Agrawal, A.J. Mohammad, Linear combination of a new sequence of linear positive operators. Rev. Un. Mat. Argent. **44**(1), 33–41 (2003)
13. A. Aral, V. Gupta, Direct estimates for Lupaş Durrmeyer operators. Filomat **30**(1), 191–199 (2016)
14. D.A. Ari, Approximation properties of Szász type operators involving Charlier polynomials. Filomat **31**(2), 479–489 (2017)
15. J.A.H. Alkemade, The second moment for the Meyer-König and Zeller operators. J. Approx. Theory **40**, 261–273 (1984)
16. H.K. Arab, M. Dehghan, M.R. Eslahchi, A new approach to improve the order of approximation of the Bernstein operators: theory and applications. Numer. Algor. **77**(1), 111–150 (2018)
17. D. Aydin, A. Aral, G. B-Tunca, A generalization of Post-Widder operators based om q-integers. Ann. Alexandru Ioan Cuza Univ. Math. **LXII**(1), 77–88 (2016)
18. K. Balázs, Approximation by Bernstein-type rational functions. Acta Math. Acad. Sci. Hungar. **26**, 123–134 (1975)

© The Author(s), under exclusive license to Springer Nature Switzerland AG 2019   91
V. Gupta and M. T. Rassias, *Moments of Linear Positive*
Operators and Approximation, SpringerBriefs in Mathematics,
https://doi.org/10.1007/978-3-030-19455-0

19. K. Balázs, J. Szabados, Approximation by Bernstein-type rational functions II. Acta Math. Acad. Sci. Hungar. **40**, 331–337 (1982)
20. V.A. Baskakov, An instance of a sequence of linear positive operators in the space of continuous functions. Dokl. Akad. Nauk SSSR **113**(2), 249–251 (1957). (In Russian)
21. B. Baxhaku, A. Berisha, The approximation Szász-Chlodowsky type operators involving Gould-Hopper type polynomials. Abstract Appl. Anal. 2017, Article ID 4013958, 8 pages (2017)
22. S.N. Bernstein, Demonstration du Théoréme de Weierstrass fondée sur le calcul des Probabilités. Comm. Soc. Math. Kharkov 2. Series XIII No. 1, 1–2 (1912)
23. W. Chen, On the modified Bernstein-Durrmeyer operators, in *Report of the Fifth Chinese Conference on Approximation Theory, Zhen Zhou, China* (1987)
24. E.W. Cheney, A. Sharma, Bernstein power series. Canad. J. Math. **16**, 241–253 (1964)
25. T.S. Chihara, *An Introduction to Orthogonal Polynomials* (Gordon and Breach, New York, 1978)
26. N. Deo, Faster rate of convergence on Srivastava-Gupta operators. Appl. Math. Comput. **218**(21), 10486–10491 (2012)
27. N. Deo, M. Dhamija, Generalized positive linear operator based on PED and IPED. Iran J. Sci. Technol. Trans. Sci. (2018). https://doi.org/10.1007/s40995-017-0477-5
28. N. Deo, M. Dhamija, D. Miclăuş, Stancu-Kantorovich operators based on inverse Pólya-Eggenberger distribution. Appl. Math. Comput. **273**, 281–289 (2016)
29. R.A. DeVore, G.G. Lorentz, *Constructive Approximation* (Springer, Berlin, 1993)
30. M.M. Derriennic, Sur l'approximation de fonctions integrables sur [0,1] par des polynomes de Bernstein modifiés. J. Approx. Theory **31**(4), 325–343 (1981)
31. M. Dhamija, N. Deo, Approximation by generalized positive linear Kantorovich operators. Filomat **31**(14), 4353–4368 (2017)
32. Z. Ditzian, V. Totik, *Moduli of Smoothness* (Springer, New York, 1987)
33. D. Dubey, V.K. Jain, Rate of approximation for integrated Szász-Mirakyan operators. Demonstratio Math. **41**, 879–886 (2008)
34. O. Dogrü, R.N. Mohapatra, M. Örkcü, Approximation properties of generalized Jain operators. Filomat **30**(9), 2359–2366 (2016)
35. J.L. Durrmeyer, Une formule d' inversion de la Transformee de Laplace, Applications a la Theorie des Moments, These de 3e Cycle, Faculte des Sciences de l' Universite de Paris 1967
36. A. Farcaş, An asymptotic formula for Jain's operators. Stud. Univ. Babeş Bolyai Math. **57**, 511–517 (2012)
37. Z. Finta, On converse approximation theorems. J. Math. Anal. Appl. **312**, 159–180 (2005)
38. Z. Finta, N.K. Govil, V. Gupta, Some results on modified Szász-Mirakjan operators. J. Math. Anal. Appl. **327**(2), 1284–1296 (2007)
39. Z. Finta, V. Gupta, Direct and inverse estimates for Phillips type operators. J. Math. Anal. Appl. **303**(2), 627–642 (2005)
40. S.G. Gal, V. Gupta, Approximation by complex Stancu Beta integral operators of second kind in semidisks. Rev. Anal. Numer. Theor. Approx. **42**(1), 21–36 (2013)
41. S.G. Gal, V. Gupta, Approximation by complex Lupas-Durrmeyer polynomials based on Polya distribution. Banach J. Math. Anal. **10**(1), 209–221 (2016)
42. S.G. Gal, V. Gupta, Approximation by complex Beta operators of first kind in strips of compact disks. Med. J. Math. **10**(1), 31–39 (2013)
43. I. Gavrea, M. Ivan, An elementary function representation of the second-order moment of the MeyerKönig and Zeller operators. Mediterr. J. Math. **15**, 20 (2018). https://doi.org/10.1007/s00009-018-1066-3
44. T.N.T. Goodman, A. Sharma, A modified Bernstein-Schoenberg operator, in *Proceedings of the Conference on Constructive Theory of Functions, Varna 1987*, ed. by Sendov et al. (Publishing House Bulgarian Academy of Sciences, Sofia, 1988), pp. 166–173
45. N.K. Govil, V. Gupta, Convergence rate for generalized Baskakov type operators. Nonlinear Anal. Theory Methods Appl. **69**(11), 3795–3801 (2008)

46. V. Gupta, A note on modified Szász operators. Bull. Inst. Math. Acad. Sin. **21**(3), 275–278 (1993)
47. V. Gupta, A note on modified Baskakov type operators. Approx. Theory Appl. **10**(3), 74–78 (1994)
48. V. Gupta, Simultaneous approximation by Szász-Durrmeyer operators. Math. Stud. **64**(1–4), 27–36 (1995)
49. V. Gupta, A note on the rate of convergence of Durrmeyer type operators for functions of bounded variation. Soochow J. Math. **23**, 115–118 (1997)
50. V. Gupta, Rate of approximation by new sequence of linear positive operators. Comput. Math. Appl. **45**(12), 1895–1904 (2003)
51. V. Gupta, A note on modified Phillips operators. Southeast Asian Bull. Math. **34**, 847–851 (2010)
52. V. Gupta, A new class of Durrmeyer operators. Adv. Stud. Contemp. Math. **23**(2), 219–224 (2013)
53. V. Gupta, Direct estimates for a new general family of Durrmeyer type operators. Boll. Un. Mate. Italiana **7**, 279–288 (2015)
54. V. Gupta, *Approximation by Generalizations of Hybrid Baskakov Type Operators Preserving Exponential Functions* (Selected Topics, Wiley, Mathematical Analysis and Applications, 2017)
55. V. Gupta, Some examples of genuine approximation operators. General Math. **26**(1–2), 3–9 (2018)
56. V. Gupta, On difference of operators with applications to Szász type operators. Revista de la Real Academia de Ciencias Exactas, Físicas y Naturales. Serie A. Matemáticas (2018). https://doi.org/10.1007/s13398-018-0605-x
57. V. Gupta, Differences of operators of Lupaş type. Constr. Math. Anal. **1**(1), 9–14 (2018)
58. V. Gupta, A.M. Acu, D.F. Sofonea, Approximation of Baskakov type Pólya Durrmeyer operators. Appl. Math. Comput. **294**, 318–331 (2017)
59. V. Gupta, A.M. Acu, On Baskakov-Szász-Mirakyan type operators preserving exponential functions. Positivity **22**(3), 919–929 (2018)
60. V. Gupta, R.P. Agarwal, *Convergence Estimates in Approximation Theory* (Springer, Cham, 2014)
61. V. Gupta, R.P. Agarwal, D.K. Verma, Approximation for a new sequence of summation-integral type operators. Adv. Math. Sci. Appl. **23**(1), 35–42 (2013)
62. V. Gupta, A. Aral, A note on Szász-Mirakyan-Kantorovich type operators preserving e^{-x}. Positivity **22**(2), 415–423 (2018)
63. V. Gupta, N. Deo, A note on improved estimations for integrated Szász-Mirakyan operators. Math. Slovaca **61**(5), 799–806 (2011)
64. V. Gupta, G.C. Greubel, Moment estimations of a new Szász-Mirakyan-Durrmeyer operators. Appl. Math Comput. **271**, 540–547 (2015)
65. V. Gupta, P. Maheshwari, Bézier variant of a new Durrmeyer type operators. Riv. Math. Univ. Parma **2**, 9–21 (2003)
66. V. Gupta, N. Malik, Th.M. Rassias, Moment generating functions and moments of linear positive operators, in *Modern Discrete Mathematics and Analysis*, ed. by N.J. Daras, Th.M. Rassias (Springer, 2017)
67. V. Gupta, Th.M. Rassias, Lupaş-Durrmeyer operators based on Polya distribution. Banach J. Math. Anal. **8**(2), 146–155 (2014)
68. V. Gupta, Th.M. Rassias, P.N. Agrawal, A.M. Acu, *Recent Advances in Constructive Approximation Theory* (Springer, Cham, 2018)
69. V. Gupta, Th.M. Rassias, J. Sinha, A survey on Durrmeyer operators, in *Contributions in Mathematics and Engineering*, ed. by P.M. Pardalos, Th.M. Rassias (Springer International Publishing Switzerland, 2016), pp. 299–312
70. V. Gupta, Th.M. Rassias, R. Yadav, Approximation by Lupaş-Beta integral operators. Appl. Math. Comput. **236**, 19–26 (2014)

71. V. Gupta, Th.M. Rassias, E. Pandey, On genuine Lupaş-Beta operators and modulus of continuity. Int. J. Nonlinear Anal. Appl. **8**(1), 23–32 (2017)
72. V. Gupta, J. Sinha, Simultaneous approximation for generalized Baskakov-Durrmeyer-type operators. Mediter. J. Math. **4**, 483–495 (2007)
73. V. Gupta, G.S. Srivastava, Simultaneous approximation by Baskakov-Szász type operators, Bull. Math. de la Soc. Sci. de Roumanie (N.S.) **37**(3-4), 73–85 (1993)
74. V. Gupta, G.S. Srivastava, Inverse theorem for Szasz Durrmeyer operators. Bull. Inst. Math. Acad Sin. **23**(2), 141–150 (1995)
75. V. Gupta, G.S. Srivastava, On convergence of derivatives by Szász-Mirakyan-Baskakov type operators. Math. Stud. **64**(1–4), 195–205 (1995)
76. V. Gupta, G.S. Srivastava, A. Sahai, On simultaneous approximation by Szász Beta operators. Soochow J. Math. **21**, 1–11 (1995)
77. V. Gupta, H.M. Srivastava, A general family of the Srivastava-Gupta operators preserving linear functions. European J. Pure Appl. Math. **11**(3), 575–579 (2018)
78. V. Gupta, G. Tachev, Approximation by Szász-Mirakyan Baskakov operators. J. Appl. Funct. Anal. **9**(3–4), 308–319 (2014)
79. V. Gupta, G. Tachev, On approximation properties of Phillips operators preserving exponential functions. Mediterr. J. Math. **14**(4), 177 (2017)
80. V. Gupta, G. Tachev, *Approximation with Positive Linear Operators and Linear Combinations*. Developments in Mathematics, vol. 50 (Springer, Cham, 2017)
81. V. Gupta, G. Tachev, Some results on Post-Widder operators preserving test function e_r, communicated
82. V. Gupta, G. Tachev, A.M. Acu, Modifed Kantorovich operators with better approximation properties. Numer. Algorithms **81**(1), 125–149 (2019)
83. V. Gupta, G. Tachev, D. Soybaş, Improved approximation on Durrmeyer type operators. Bol. Soc. Mat. Mex. (2018). https://doi.org/10.1007/s40590-018-0196-8
84. V. Gupta, D.K. Verma, P.N. Agrawal, Simultaneous approximation by certain Baskakov Durrmeyer Stancu operators. J. Egypt. Math. Soc. **20**(3), 183–187 (2012)
85. V. Gupta, R. Yadav, Direct estimates in simultaneous approximation for BBS operators. Appl. Math. Comput. **218**(22), 11290–11296 (2012)
86. A. Holhoş, The rate of approximation of functions in an infinite interval by positive linear operators. Stud. Univ. Babeş-Bolyai, Math. (2), 133–142 (2010)
87. R.C. İlbey, Weighted approximation properties of Dunkl analogue of modified Szász-Mirakjan operators. Int. J. Math. Comput. Methods, http://www.iaras.org/iaras/journals/ijmcm
88. N. Ispir, I. Yuksel, On the Bézier variant of Srivastava-Gupta operators. Appl. Math. E Notes **5**, 129–137 (2005)
89. N. Ispir, C. Atakut, Approximation by modified Szász-Mirakjan operators on weighted spaces. Proc. Math. Sci. **112**(4), 571–578 (2002). (Indian Academy of Sciences)
90. M. Ismail, C.P. May, On a family of approximation operators. J. Math. Anal. Appl. **63**, 446–462 (1978)
91. G.C. Jain, Approximation of functions by a new class of linear operators. J. Austral. Math. Soc. **13**, 271–276 (1972)
92. G.C. Jain, S. Pethe, On the generalizations of Bernstein and Szász-Mirakjan operators. Nanta Math. **10**, 185–193 (1977)
93. A. Jakimovski, D. Leviatan, Generalized Szász operators for the approximation in the infinite interval. Mathematica (Cluj) **11**, 97–103 (1969)
94. H.S. Kasana, G. Prasad, P.N. Agrawal, A. Sahai, Modified Szász, operators, in *Conference on Mathematical Analysis and its Applications*, Kuwait, vol. 1988 (Pergamon Press. Oxford, 1985), pp. 29–41
95. H. Karsali, Rate of convergence of a new Gamma type operators for functions with derivatives of bounded variation. Math. Comput. Model. **45**(5–6), 617–624 (2007)
96. H. Karsali, V. Gupta, Some approximation properties of q-Chlodowsky operators. Appl. Math. Comput. **195**, 220–229 (2008)

97. M.K. Khan, Approximation properties of Beta operators, in *Progress in Approximation Theory* (Academic Press, New York, 1991), pp. 483–495
98. G.G. Lorentz, *Bernstein Polynomials* (University of Toronto Press, 1953)
99. A. Lupaş, Die Folge der Betaoperatoren, Dissertation, University Stuttgart, 1972
100. A. Lupaş, The approximation by means of some linear positive operators, in *Approximation Theory, Proceedings of the International Doortmund Meeting IDoMAT 95, Witten, Germany, March 13–17, Mathematical Research*, ed. by M.W. Muller, M. Felten, D.H. Mache, vol. 86 (Akademie Verlag, Berlin, 1995), pp. 201–229
101. L. Lupaş, A. Lupaş, Polynomials of binomial type and approximation operators. Stud. Univ. Babeş-Bolyai Math. **32**(4), 61–69 (1987)
102. A. Lupaş, M. Müller, Approximationseigenshaften der Gammaoperatoren. Math. Z. **98**, 208–226 (1967)
103. C.P. May, Saturation and inverse theorems for combinations of a class of exponential-type operators. Canad. J. Math. **28**, 1224–1250 (1976)
104. C.P. May, On Phillips operator. J. Approx. Theory **20**(4), 315–332 (1977)
105. S.M. Mazhar, V. Totik, Approximation by modified Szász operators. Acta Sci. Math. (Szeged) **49**, 257–269 (1985)
106. V. Miheşan, Gamma approximating operators. Creat. Math. Inf. **17**, 466–472 (2008)
107. G.M. Mirakjan, Approximation des fonctions continues au moyen de polynomes de la forme $e^{-nx} \sum_{k=0}^{m_n} C_{k,n} x^k$ [Approximation of continuous functions with the aid of polynomials of the form $e^{-nx} \sum_{k=0}^{m_n} C_{k,n} x^k$]. Comptes rendus de lÁcadémie des sciences de lÚRSS **31**, 201–205 (1941). (in French)
108. W. Meyer-König, K. Zeller, Bernsteinsche Potenzreihen. Stud. Math. **9**, 89–94 (1960)
109. D. Miclăuş, The revision of some results for Bernstein-Stancu type operators. Carpathian J. Math. **28**(2), 289–300 (2012)
110. D. Miclăuş, The moments of Bernstein-Stancu operators revisited. Mathematica **54**(77), no. 1, 78–83 (2012)
111. V. Miheşan, Uniform approximation with positive linear operators generated by generalized Baskakov method. Automat. Comput. Appl. Math. **7**(1), 34–37 (1998)
112. D.C.-Morales, V. Gupta, Two families of Bernstein-Durrmeyer type operators. Appl. Math. Comput. **248**, 342–353 (2014)
113. Lopez-Moreno, A.-J., Latorre-Palacios, J.-M.: Localization results for generalized Baskakov/Mastroianni and composite operators. J. Math. Anal. Appl. **380**(2), 425–439 (2011)
114. R. Păltănea, Modified Szász-Mirakjan operators of integral form. Carpathian J. Math. **24**, 378–385 (2008)
115. R. Păltănea, Estimates of approximation in terms of a weighted modulus of continuity. Bull. Transilvania Univ. Brasov **4**(53), 67–74 (2011)
116. G.M. Phillips, *Interpolation and Approximation by Polynomials* (Springer, New York, 2003)
117. R.S. Phillips, An inversion formula and semi-groups of linear operators. Ann. Math. **59**, 325–356 (1954)
118. O.T. Pop, M. Farcaş, About Bernstein polynomial and Stirling's numbers of second kind. Creative Math. **14**, 53–56 (2005)
119. T. Popoviciu, Sur l'approximation des fonctions covexes d'ordre supérieur. Mathematica (Cluj) **10**, 49–54 (1934)
120. T. Popoviciu, Sur l'approximation des fonctions continues d'une variable réelle par des polynomes. Ann. Sci. Univ. Iasi. Sect. I Math. **28**, 208 (1942)
121. G. Prasad, P.N. Agrawal, H.S. Kasana, Approximation of functions on $[0, \infty]$ by a new sequence of modified Szász operators. Math. Forum **VI**(2), 1–11 (1983)
122. R. Pratap, N. Deo, Approximation by genuine Gupta-Srivastava operators. Revista de la Real Academia de Ciencias Exactas, Físicas y Naturales. Serie A. Matemáticas (2019). https://doi.org/10.1007/s13398-019-00633-4
123. R.K.S. Rathore, Linear combinations of linear positive operators and generating relations on special functions, Ph.D. thesis, Delhi, 1973

124. R.K.S. Rathore, O.P. Singh, On convergence of derivatives of Post-Widder operators. Indian J. Pure Appl. Math. **11**, 547–561 (1980)
125. L. Rempulska, M. Skorupka, On approximation by Post-Widder and Stancu operators preserving x^2. Kyungpook Math. J. **49**, 57–65 (2009)
126. M. Rosenblum, Generalized Hermite polynomials and the Bose-like oscillator calculus. Oper. Theory Adv. Appl. **73**, 369–396 (1994)
127. A. Sahai, G. Prasad, On simultaneous approximation by modified Lupas operators. J. Approx. Theory **45**(2), 122–128 (1985)
128. R.P. Sinha, P.N. Agrawal, V. Gupta, On simultaneous approximation by modified Baskakov operators. Bull. Soc. Math. Belg. Ser. B **43**(2), 217–231 (1991)
129. H.M. Srivastava, V. Gupta, A certain family of summation-integral type operators. Math. Comput. Model. **37**, 1307–1315 (2003)
130. D.D. Stancu, Approximation of functions by a new class of linear polynomial operators. Rew. Roum. Math. Pure. Appl. **13**, 1173–1194 (1968)
131. D.D. Stancu, Two classes of positive linear operators. Anal. Univ. Timisora Ser. Matem. **8**, 213–220 (1970)
132. S. Sucu, Dunkl analogue of Szász operators. Appl. Math. Comput. **244**, 42–48 (2014)
133. D.D. Stancu, On the beta approximating operators of second kind. Rev. Anal. Numér. Théor. Approx. **2**(4), 231–239 (1995)
134. O. Szász, Generalizations of S. Bernstein's polynomials to the infinite interval. J. Res. Natl. Bur. Stand. **45**(3), 239–245 (1950)
135. G. Tachev, V. Gupta, General form of Voronovskaja's theorem in terms of weighted modulus of continuity. Results Math. **69**(3–4), 419–430 (2016)
136. S. Tarabie, On Jain-Beta linear operators. Appl. Math. Inf. Sci. **6**(2), 213–216 (2012)
137. S. Varma, F. Taşdelen, Szász type operators involving Charlier polynomials. Math. Comput. Model. **56**, 118–122 (2012)
138. S. Varma, S. Sucu, G. İçöz, Generalization of Szász operators involving Brenke type polynomials. Comp. Math. Appl. **64**, 121–127 (2012)
139. E. Voronovskaja, Determination de la forme asymptotique d'approximation des fonctions par les polynomes de M. Bernstein. Dokl. Akad. Nauk SSSR **4**, 86–92 (1932)
140. D.V. Widder, *The Laplace Transform* (Princeton University Press, Princeton, N.J, Princeton Mathematical Series, 1941)
141. X.M. Zeng, Approximation properties of Gamma operators. J. Math. Anal. Appl. **311**, 389–401 (2005)